A History of the Regular Mounted Infantry

A History of the Regular Mounted Infantry

From Victoria's Colonial Wars to the Russian Campaign of 1919

Paul Baker

Pen & Sword
MILITARY
AN IMPRINT OF PEN & SWORD BOOKS LTD.
YORKSHIRE – PHILADELPHIA

First published in Great Britain in 2024 by
Pen & Sword Military
An imprint of
Pen & Sword Books Ltd
Yorkshire - Philadelphia

ISBN 978 1 03610 651 5

A CIP catalogue record for this book is available from the British Library.

Typeset in INDIA by IMPEC eSolutions
Printed and bound in England by CPI (UK) Ltd.

Pen & Sword Books Limited incorporates the imprints of Archaeology, Atlas, Aviation, Battleground, Digital, Discovery, Family History, Fiction, History, Local, Local History, Maritime, Military, Military Classics, Politics, Select, Transport, True Crime, After the Battle, Air World, Claymore Press, Frontline Publishing, Leo Cooper, Remember When, Seaforth Publishing, The Praetorian Press, Wharncliffe Books, Wharncliffe Local History, Wharncliffe Transport, Wharncliffe True Crime and White Owl.

For a complete list of Pen & Sword titles please contact

PEN & SWORD BOOKS LIMITED
47 Church Street, Barnsley, South Yorkshire, S70 2AS, England
E-mail: enquiries@pen-and-sword.co.uk
Website: www.pen-and-sword.co.uk

or

PEN AND SWORD BOOKS
1950 Lawrence Rd, Havertown, PA 19083, USA
E-mail: uspen-and-sword@casematepublishers.com
Website: www.penandswordbooks.com

MIX
Paper | Supporting responsible forestry
FSC www.fsc.org
FSC® C013604

Contents

Preface

An online auction offered for sale an unpublished, incomplete manuscript intending to be a short history of the mounted infantry for remaining members of the Mounted Infantry Club. Included were various letters and first-hand accounts resulting from an appeal by Major Maurice Tomlin[1] who, in July 1936, sent out on behalf of the President of the Club, General Sir Ian Hamilton[2], requests to battalions, depots and editors of regimental journals for records of services and war diaries which dealt with the mounted infantry. He also requested them to notify serving and retired officers that their personal reflections were being sought.

Major Tomlin, recognising problem of aging membership, had written to senior members of the MI Club on 30 March 1936 stating that the club was like a tontine (an investment linked to a living person which provides an income for as long as that person is alive). Since the disbandment of the mounted infantry on the eve of the First World War the only new members were the

---

1 **Major Maurice Hilliard Tomlin OBE.** Born 28 June 1868, died 10 August 1945. Commissioned into the Yorkshire Regiment (Alexandra, Princess of Wales's Own Yorkshire Regiment) in 1888. Served in operations on the Niger 1897-98, Boer War commanding Northern Company, 2nd MI Battalion, in 1901. Left the army in 1908 and joined the police force where he was appointed a Chief Constable in the Metropolitan Police in 1912 and Assistant Commissioner in 1932.

2 **General Sir Ian Hamilton GCB, GCMG, DSO, TD.** Born 16 January 1853, died 12 October 1947. In 1900 commanded a mounted infantry division and ended the war in 1902 commanding the military columns operating in the Western Transvaal. In 1905 he was military attaché for the British Indian Army with the Japanese forces serving in Manchuria during the Russo Japanese war. He went on to command the Mediterranean Expeditionary force in the ill-fated Gallipoli campaign of 1915. He was President of the Mounted Infantry Club.

winners of their races at staff college point-to-point meetings. He proposed that new membership should be extended to all infantry officer winners of point-to-point races. However, this would not be sufficient to prolong the club's existence as aging membership and the forthcoming Second World War put paid to its long-term survival.

The appeal elicited several responses and Major Tomlin worked on the manuscript, which appeared to have been completed in 1939 though, by the outbreak of the war, it had not been published. Tragically, it appears that during a wastepaper salvage drive at the start of the war an enthusiastic cull of papers by Tomlin's family included much of the work he had completed for this project. This would have been the end of the project but, in August 1942, Brigadier Standish G. Craufurd[3] wrote to Tomlin to see what was happening regarding publication of the manuscript. When he realised nothing was happening and most of the original manuscript lost, he undertook to rewrite the work using what material remained and his own resources. His view was, 'That it is a great pity to let slip an account of a side issue of the old Army, a very distinct phase, just before the great change brought by the Great War.' He went on to say that, 'If the publication is much delayed it will be too late for the majority of the participants.'

With Craufurd completing his draft by mid-1943 and the final input from General Sir Alick Godley[4], Sir Ian Hamilton (then 91) was approached to write

---

[3] **Brigadier General Sir George Standish Gage Craufurd, CB, CMG, CIE, DSO.** Born 10 November 1872, died 6 January 1956. Commissioned into the Gordon Highlanders in 1892, served in India, with the 6th Mounted Infantry in the Boer War and ended the First World War as Temporary Brigadier General commanding 18th Infantry Brigade on the Western Front.

[4] **General Sir Alexander Godley GCB, KCMG.** Born 4 February 1867, died 6 March 1957. Appointed adjutant of the MI School at Aldershot in 1895 and Alderson's adjutant for the MI Battalion that was sent out to South Africa in 1896 to quell the rebellion originating in Matebeleland and spread to Mashonaland. He served in the Boer War helping to raise irregular mounted units, serving as Baden Powell's adjutant during the siege at Mafeking , as Plumer's Chief Staff officer and then commanding the Rhodesian Brigade. Godley's talent for training was recognised by Field Marshal Roberts and in 1901 he was appointed DAAG on the

the foreword. His response was not supportive; he felt this was a 'makeshift history' that would best be left until the end of the war, when access to records would be easier and the manuscript could be improved upon. Thus, the work remained in the hands of Major Tomlin in its unpublished state. He passed away in August 1945 as the war was ending, followed by Sir Ian Hamilton in 1947 and Craufurd in 1956.

Despite Sir Ian Hamilton's opinion, the manuscript and correspondence provide a unique and interesting archive by surviving members of the mounted infantry. Considering that history of this corps has been poorly documented, this collection of reminiscences and research by mounted infantry veterans deserves a wider audience.

HQ staff at Aldershot, which carried with it command of the School of Instruction for mounted infantry. In 1903 he commanded the Mounted Infantry School at Longmoor. In the First World War he was the Divisional Commander of the New Zealand and Australian Division serving in the ANZAC Corps at Gallipoli and replaced Birdwood when he succeeded Hamilton in Command of the Dardanelles Army towards the end of the campaign. Commanded II ANZAC when they were moved to France. He ended the war commanding XXII Corps. In 1923 he was promoted General and in 1928 he was given the governorship of Gibraltar.

Introduction

Brigadier General Sir George Standish Gage Craufurd, who served with the 6th Mounted Infantry Battalion during the Boer War, wrote, 'Regarding the MI history, I feel it was a very distinct phase in the old British Army. It was, in fact, a dying flash of the old Regular Army which was almost annihilated in 1914/15 and rose from its ashes a new creation albeit, with old tradition.' He penned this in a letter to Major Maurice Tomlin in September 1942 when the two were involved in writing a short history of the mounted infantry intended for the members of the MI Club, a club formed of surviving veterans who had served in the mounted infantry. Although that manuscript was not published, the draft, original correspondence and recorded reminiscences of members have survived. As a collection, these eyewitness accounts provide an insight into how the Victorian Army met the requirement for mobile forces in an age before the internal combustion engine and today's modern mechanised units - the successor to the mounted infantry. The principal role of the mounted infantry was:

a) The rapid transportation of a fighting force to a distant point, where they would fight dismounted.
b) Reconnaissance over a wide expanse of country, beyond the capacity of infantry.

The mounted infantry had a short, but active, existence playing a significant role in the British Army's small wars of the Victorian period, a major role in the war in South Africa and continued as an influence up until its disbandment on the eve of the First World War. Although, as will be seen in the last account, the mounted infantry concept could be resurrected as the need arose and perform a role as evidenced in this account of the British expedition to Russia in 1919.

With virtually all battalions having representation in the MI battalions at some point during the South Africa war of 1899-1902, accounts and records are widely disbursed and have not been centralised in an orderly, accessible manner. What records exist are found in personal papers and regimental records that generally focus on their participation in MI operations. Published records tend to be found in biographies, regimental journals and histories. The lack of centralised records can be explained by the fact that the MI was not a separate, permanent unit, although from 1888 it was responsible for the training of the mounted infantry. Following their training period soldiers would return to their parent units to be abstracted to form MI units as required when hostilities commenced. Mounted infantry battalions were therefore impermanent units composed of trained infantry formed into sections or, as the South African war progressed, as complete companies, when required and returning to their parent units at the end of operations. An example of the composition of 9th Battalion Mounted Infantry in January 1901 illustrates the various battalions represented in the single MI battalion:

No. 1 Company	2nd Battalion Derbyshire Regiment
Malta Mounted Infantry	2nd Battalion Loyal North Lancashire Regiment
	2nd Battalion Royal Warwickshire Regiment
	3rd Battalion Lancashire Fusiliers
No. 2 Company	1st Battalion Derbyshire Regiment
No. 3 Company	2nd Battalion Royal Irish Rifles
No. 4 Company	1st Battalion Royal Munster Fusiliers

The Boer War saw expediential growth of the MI from the 1st and 2nd Mounted Infantry battalions formed at the MI school, Aldershot that sailed to South Africa with the Cavalry Division, and the immediate expansion to eight battalions (by requiring infantry battalions already serving South Africa to form a mounted infantry company). Further battalions were added during the course of the war so that by its end about 16,000 soldiers had served in the MI.

With virtually all battalions being represented at some point in the war, surviving accounts and records were widely disbursed. Brigadier Craufurd served with the Gordon Highlanders Company of the 6th Mounted Infantry and wrote an account of their exploits originally intended for their regimental history. This account is used here as a representative example of MI operations during the Boer War. The company served from the date of its formation to the day peace was declared, as a unit continuously engaged on column work in different areas including the Transvaal, Free State and Cape Colony. Veterans from other battalions provided their memoirs, but space here prevents their inclusion.

Tomlin and Craufurd were successful in obtaining accounts from members for small colonial conflicts in Burma and Northern Nigeria involving the use of mounted infantry with their own peculiarities. Burma used sturdy little Burmese ponies averaging from 13 hands to 13.2 hands, while the Nigerian MI was formed from specially enlisted locals and was the only MI unit that did not rely on abstraction to provide its complement.

Their close relations with General Sir Alexander John Godley elicited his contribution of additional notes from the Mounted Infantry Schools of Instruction, Aldershot where he was attached for instruction in 1894. Fellow officers included Thomas D'Oyly Snow, Henry Wilson, and Harry Rawlinson of the Coldstream Guards. He was appointed commandant in 1901 and published his autobiography, *Life of an Irish Soldier*. He also suggested using the chapter he contributed to E.A.H. Alderson's *With the Mounted Infantry and the Mashonaland Field Force, 1896*, detailing his experience in the expedition.

The first-hand accounts brought together in this book have either been specially written by veterans of the Victorian conflicts in response to the original appeal for information sent out in 1936 or have been selected by Tomlin and Craufurd from lesser-known published sources. The accounts written by Victorian soldiers during the period 1936–1943 reflect the thoughts and ideas of the late nineteenth and early twentieth century, and are reproduced here as they wrote them, with minor grammatical or spelling amendments.

Chapter 1

Early history

Mounted infantry, as a recognised arm to the British Regular Army, had a meteoric, but brief, career in the last quarter of the nineteenth century. It reached its zenith in the South African war only to sink into decline with the reorganisation of the Army prior to the First World War. That quarter of a century coincided with a great expansion of the British Empire in the latter half of Queen Victoria's reign.

Mounted infantry, in one form or another, is, however, a constantly recurring phase of military history stretching back to the earliest campaigns on record. Victory on the field of battle generally comes to those who can put in a smashing blow at the critical moment. Hence the problem of transporting the archer rapidly to the point where his firepower could be most effective must always have exercised the mind of every military commander. The chariot offered the earliest solution - rapidity of transport, freedom of action, accuracy of aim, and weight behind the attack. This was in use towards the close of twentieth century BC with the Pharaoh, King of Egypt employing his chariots and archers in the Libyan desert to repel an invasion of 'blond barbarians', threatening the civilization of his day in much the same way as General Alexander employed his mobile infantry columns to repel another invasion of blond barbarians threatening the same historic land of Egypt in the twentieth century AD.

The military genius of Alexander the Great and his natural inclination and aptitude for the cavalry branch is to whom we owe the earliest instance of the cavalrymen trained as a dragoon. In the early days of the Roman Republic, the Roman cavalry seemed to have been kept in reserve, either for pursuit or to reinforce a critical point on the field of battle. In this latter case, they fought dismounted, thus performing the duties of mounted infantry rather than true cavalry. The Germanic tribes employed as auxiliaries by the Romans fought, at first, entirely as mounted infantry. They used their small horses principally to give them tactical mobility on the field of battle. In later campaigns they were better mounted by their Roman employers and trained in true cavalry tactics and the idea of mounted infantry vanished from the field.

The fall of Rome was also to witness the revival of the East under the followers of Mohammed, and under the impact of Islam, the Franks, as the nations of Western Europe were now called, were to regain much lost knowledge.

Feudalism became a highly organised institution. After the Crusades, chain armour gradually gave way to plate armour, and for a century the knight in armour reigned supreme. In their coats of mail they were almost unassailable by the lighter-armed feudal infantry, mostly conscripted serfs.

During the French Wars of the fourteenth century, the English longbow archers at Crecy, at Poitiers and at Agincourt were able to defeat the mailed chivalry of France. At Poitiers the timely use of the mounted English archers contributed in no small measure to the final victory and was to form a precedent for future campaigns. The English longbowmen too had not unworthy competitors in the Swiss Free Companies of mercenaries with their powerful, if slower, crossbows. In any event the bowman could no longer be ignored and the mounted archer, known thereafter as hobelars, had come to stay and each nation in turn adopted them. From the hobelar of the fifteenth came the arquebusier of the sixteenth, the carabiner of the seventeenth, and the dragoons of the eighteenth, to be succeeded by the mounted rifleman and mounted infantry of the nineteenth century. These, in turn, gave place to the mechanised forces of today.

Gunpowder was to bring another medium-range weapon into the field with portable firearms first mentioned among the Flemings around the middle of the fourteenth century. By the beginning of the fifteenth century ballistics had sufficiently improved for lead balls fired by a small hand cannon to penetrate the solid armour, and the petronel, the first crude attempt at a firearm, came into use.

These petronels (so called because they were rested against the poitrail or cuirass when fired) were plain tubes with a straight wooden stock and were fired with a match applied to a touch-hole at the top. The end of the stock rested against the breast, while the muzzle was steadied by arrest with a fork or crutch which stood up from the pummel of the saddle. Here we had the mounted rifleman in embryo. Then came improvements in the stock, enabling it to be fired from the shoulder with a touch-hole on the right side. There followed further improvements till about the latter end of the fifteenth century, the arquebus was invented, which was supplied with a matchlock, a trigger, and a spring. This was a very good weapon and the prototype of the rifle of today. The wheellock was next, invented in 1515 at Nuremberg, but did not entirely replace the matchlock, which was simpler, surer and more reliable.

The greater the progress in the application of gunpowder to military purpose, the more the horsemen desired to make use of it and apply it to their own service, resulting in the rapid spread of firearms in the cavalry service all over Europe. By the time of Louis XIII, the French cavalry was organised almost entirely on mounted rifleman and the actual fighting was done on foot. The idea of mounting archers in order to save them from fatigue had long been practised in the French Army, as a temporary expedient, but the French were the first to revive the use of dragoons as a special branch of the cavalry in the fighting in Piedmont under Marshall De Brissac in 1550. Companies of arquebusiers were mounted on horseback to enable them to move more rapidly into position where their presence was required and where they would be obliged to dismount to fight. The objective being to supply increased mobility to certain portions of his infantry. This was the real idea and object of dragoons and seems to have been the first systematic use of this type of soldier in modern times.

The dragoons then, in the first place, were simply infantry and the horse was used solely for rapidity of movement, the fighting being always done on foot. This gradually changed, the men being averse to dismount, so that they soon became accustomed to fighting more on their horses than on foot, which after time led to this arm falling out of favour.

Meantime, from 1611 onwards during the Thirty Years War, dragoons were conspicuously exploited by the German states. He was regarded wholly as an infantryman, was paid at the same rates of pay and was punished under the Infantry Code. It is here that we encounter the first record of mounted infantry in their proper sphere, for during Wallanstein's arduous marches in the Thirty Years War his dragoons with their "muskets, shovels and matchlocks" were sent on to secure the passes.

When Louis XIV came to the throne, he found only one regiment of dragoons in the French Army. In 1659 he initiated a second and, by 1690, the total force of this kind in France had increased to forty-three regiments. In the main this increase was due to the fact that in long wars the dragoons were found more suitable, as they had greater solidity than other arms and were better able to endure fatigue. Napoleon imitated the idea in Spain for the same reason. It was during the latter half the seventeenth century that dragoons were used to the greatest extent. In fact, almost all the cavalry were virtually dragoons and were often used on foot when no infantry were to hand for the purpose.

Unfortunately, the rapid development of the cavalry arm made it clear that mounted infantry were too clumsy for use in pitched battle where the ground was of a favourable nature. Light cavalry, such as that used by Prince Rupert in the Civil War, could run rings round them in the open country and cut them to pieces before they had time to reload their muskets. Consequently, the dragoons came to rely more and more on the lance and sword and finally became merged with the cavalry altogether.

Marlborough, profiting by the lessons of the Civil War, began once again to use cavalry in shock action. Hitherto the power of the musket had been the overriding factor. Marlborough had not, however, fully grasped the power of speed in shock tactics.

It was not, however, until the time of Frederick the Great that cavalry were once again trained to rely on the *arme blanche* shock tactics to achieve the objective - a mode of operation which lasted until the dominance of magazine rifle and machine-gun fire.

Frederick's first charge in his cavalry training was to prohibit absolutely the use of firearms when mounted, and to rely upon the charge at full speed, sword in hand. He told his horsemen to disregard the fire of his enemy squadrons and to close with the utmost vigour. He was also careful to train them to preserve close order and correct alignment over long distances. Cavalry, for the moment, was at its zenith. For another half century the well-directed cavalry charge could carry all before it. Yet, at Waterloo, British infantry were to stand unbroken against the finest cavalry that Napoleon could send against them.

Though dragoons had become merged in the cavalry by the end of the seventeenth century as far as all the western nations were concerned, there was a reassessment of their use in their original role in Russia under Peter the Great in his wars with Charles XII of Sweden. When Peter ascended the throne, the army consisted of two dragoon regiments and twenty-seven infantry. Peter, perceiving the extreme weakness of his cavalry, began at once to increase his dragoons. In 1701 he added another eight regiments and more in the next few years, so that by 1711 he had thirty regiments and three regiments of dragoon grenadiers.

The dragoons were organised for the express purpose of fighting both on foot and on horseback, their arms and equipment being specially designed for it. They had muskets with which they fought when dismounted, long straight swords and pistols for use when fighting as cavalry on horseback. The intention of using them principally as dragoons fighting on foot is clearly shown by the fact that 1/5th of his men were issued with axes, 1/10th with large shovels, and 1/10th of them with spades, all carried in the same way attached to the saddles. These implements were to enable them to prepare abatises and throw up field entrenchments for cover when using their muskets in defending positions.

In September 1708, Lewenhaupt was endeavouring to bring an important convoy to Charles XII. He was attacked by the Russians and fell back on Lesnaya to give battle under more favourable conditions. Czar Peter I followed closely behind. To hide his advance, he dispatched a column of dragoons, 1,000 strong under Colonel Campbell, and prepared his attack. His column consisted of an infantry regiment, a dragoon regiment and two Guards regiments. Owing to the nature of the ground, he dismounted his dragoons to fight as infantry. After a very heavy fight the Swedish advanced guard was driven back on their main body. Under cover of darkness the Swedes retreated leaving their baggage behind them. On 29 September Lewenhaupt abandoned and burnt his convoy after again being pursued by the dragoons and irregular horse who, once more, defeated and drove in his rearguards. The loss of this convoy, with an immense amount of supplies, proved an irreparable loss to the Swedish army. The final defeat of Charles XII at Pultoua followed within a year.

Though Marlborough reintroduced shock tactics into the British Army, there was a sudden revival of purely mounted infantry tactics during Lord Peterborough's fighting in Spain and upon this occasion the tactics employed and the strange manner in which the affair arose coincided more evenly with the conceptions of mounted infantry as used in its heyday from the 1880s onwards.

John Freind accompanied Lord Peterborough on his Spanish expedition, in the capacity of physician to the army, and published his account of the use of mounted infantry under the title of *Being an account of Earl Peterborough's conduct in Spain since the raising of the siege of Barcelona, and the Campagne of Valencia.*

Dated January 1705/06, he wrote: 'Arrived at Castilion de la Plama, a town rich, populous, though in the enemy's hands well affected and abounding in horses. Lord Peterborough decided that it was necessary to drive the enemy from Nules, a town three days march from Valencia, and one from Castillon.

'Nules was the most affected town in all the Kingdom to the interests of King Charles (of Spain); and fortified with the most Regular Towers, with the best walls, and in the best repair. In addition it was necessary to possess some other little Wall'd towns, which covered Castillon, such as Villa Real and Barriana.

'The enemy marched away from Nules and left a thousand of the townspeople well-armed in it. There was little hope that his lordship should gain admittance there but the enemy having the day before, partly by force, partly by treaty, got into Villa Real, and under pretence of their residence, put them all to the sword, his lordship flattered himself that terror might prevail on the inhabitants of Nules, and took a resolution as bold as fortunate. He rode in person with his Party of Horse up to the Gates of the town, under the shot of all the musquets, and required them immediately to send him out their Chief Magistrate or a Priest, if they would avoid being put to the sword without mercy, as soon as the Artillery or Mines could make a breach.

'Some priests, who knew the Earl of Peterborough, came out; my Lord, to improve their frights, allowed them but six minutes time to return with an answer. The Gates were immediately opened: and here the Earl of Peterborough began to form that cavalry which secur'd the Kingdom of Valencia, and far tributed to save Barcelona; taking near 200 Horse in this place where the enemy an hour before had 3,000 men.

'After this action, which forc'd the Duke of Anjou's troops to a second march the same day, to get further from so terrible an enemy, his lordship made a short turn that came to Castillon de la Plana.

'In this considerable town and some of the adjacent places, his lordship procured and bought near 800 horses, under the enemy's nose: persuading the whole country that his Army was now following the enemy driving them out of the Kingdom, and that some of his troops march back to take possession of the horses he had now provided. The dispatch, the persuasions, the arts used on this occasion are not to be conceived: the whole had at the time a very Romantick air, tho' every particular was before prepared and advised by his lordship.'

This was the first occasion when infantry soldiers were definitely mounted for use as horse.

Freind continued: 'I believe was equal to that of the officers and soldiers of Colonel Pierce's Regiment, who had orders to march from Vinoros, where they were sent with the rest of the 900 English infantry, to a place called Oropesa, four leagues from Castillon de la Plana. In this place, by ten in the morning, they were met by the Earl of Peterborough in a plain just bordering on the town. His lordship having made review, was congratulating the Regiment, and wishing he had but Horses and Accoutrements to try whether a corps of so good a character would maintain the like reputation upon such a charge. They, no doubt concurred very heartily with his lordship in his wishes; little expecting the execution of them in a moment. But his lordship having ordered his Secretary to give the Commissions already prepared, the officers at last believed the General in earnest when glancing to the edge of the hill they saw eight bodies of Horse drawn up separately, and found them all readily accoutred. Among these there were three good horses for each Captain, two for each Lieutenant and one for the Cornet.

'My Lord left to the field officers the choice of their troops, the other captains drew lots; and immediately they all mounted and marched to the Quarters appointed for them.'

So that conjuring trick, as it must have seemed, came off well and the victims were as pleased as the conjurer. In later days it was not quite so easy to perform the trick, possibly because the bulk of the soldiers in the eighteenth century were country-bred and accustomed to horses; they were thus able to take over their mount and walk-march straight off. Having thus added to his mounted troops, Lord Peterborough used them to give the effect of numbers by their mobility and in the event came to employ them as horse by shock action and not as dragoons only.

Chapter 2

Early Victorian Colonial campaigns

During Queen Victoria's reign the Empire was constantly expanding, and, in the process, many small wars took place. In several of these, mounted infantry played a useful part.

The ideal situation was for military expeditions to be accompanied by a body of mounted troops whether it was cavalry, mounted infantry or irregulars. Limitations in the availability of cavalry to accompany expeditions in the colonies led the local commanders to resort to mounting available troops to perform the role; however, these makeshift forces often lacked equipment and training to function efficiently in a mounted role. This transformation was not immediate nor indeed so easy. A lot of accounts of the formation of mounted infantry sections and companies give much space to the difficulties in teaching the rudiments of horsemanship and an account of a mounted infantry company at the Battle of Giniss on the Nile in 1886 ends with the statement: '1 man, a Private in the 42nd was wounded. No one fell off.'

Just before Queen Victoria was crowned, the Sixth Cape Frontier War (a series of nine wars fought between Dutch and British settlers of the eastern Cape and the Xhosa Kingdom) broke out. This followed Xhosa cattle raids and the killing, in December 1834, of a Xhosa chief by the settlers, which culminated in the dead chief's brother, Magoma, invading Cape Colony with an army of 10,000 warriors. The regular force on the frontier amounted to fewer than 800 men. The British and Dutch settlers took to safety in Grahamstown, forming a committee to oversee the erection of what turned out to be faulty defences. The following account is taken from the *Times*, dated 16 January 1935, by Sir John Fortescue: 'The British Governor, Sir Benjamin D'Urban, a veteran of the Peninsula, sent his chief staff officer, Colonel Harry Smith, with all speed to the scene of the action. Harry Smith, a man not known to make the least of his own achievements, is supposed to have covered the 600 miles to Grahamstown in six days. Moreover, upon dismounting he did not retire to rest, but took command at once, organised the settlers into troops and companies and completely restored their confidence. Then, arguing that nothing would make the Xhosas retire so quickly as invasion of their own territory, he boldly took the offensive at once with two small columns and forced two powerful chiefs to fly for their lives.

'Simultaneously, he began the work of raising native levies, and finding the Hottentots "almost peerless in their natural aptitude for military training", he formed some 800 of them into two battalions. They included, as he admitted, "every loose vagabond that he could lay his hands on," but they were turned into good soldiers in an incredibly short time. He called out also some 3,500 burghers, who came up willingly enough, but generally bought with them, on an average, one wagon to every three men, though every man somehow lacked a shirt or shoes or saddle or bridle. Comfort was their first thought in a campaign, which they never expected to last for more than a few days.

'Smith also mounted fifty men of the 75th (1st Gordon Highlanders), who took to their new calling with enthusiasm. By the end of March, he had a mobile force of 3,000 men, two thirds of them mounted, with six guns, and the stationary force on the frontier of some 2,000 more.

'They must have been in outward appearance the most astonishing British soldiers of their time. At home the traditions of George IV, the tailor, were still strong, and military uniforms were hopelessly unfitted for active service, especially in wild country. The mounted men of the 75th had abjured their trews for brown buckskin trousers and wore large red leather peaks over their forage caps, which made every man "look exactly as if he had sore eyes" - a gruesome spectacle. The Boers had huge white slouched hats, slovenly clothes, and wretched saddlery. The Hottentots had little low-crowned hats, black jackets and trousers. The Cape Mounted Rifles were in shabby green, much the worse for wear; and native guides and the like were in their local costume. Such a gang of ruffians has rarely been seen even in the annals of British warfare.

'But, under Harry Smith, they were first rate men for their work. D'Urban, an old staff officer of the Peninsula, entertained old fashioned notions about concentrating on the frontier, and advancing to fight a pitched battle. Harry Smith had as fine a record of service as D'Urban, but, knowing that cattle were the Xhosa's dearest possession, turned the war into a series of cattle raiding forays. "You gallop in and half by force, half by stratagem, burn their homes, lift their cattle and return home triumphant." His patrols were everywhere, but he made it a rule that they should observe the utmost vigilance and never move beyond reasonable distance of support. If any considerable body of Xhosas could be located in any particular spot, he attacked it as speedily as possible with the first troops he could collect. There was, he admitted, some risk in the proceedings, but, as he added, "the great science of war is to adapt its principles to the enemy that you had to contend with and to the nature of the country."

'His columns varied in strength from 100 to 500 men, but generally fell below the latter figure. The men soon became madly keen after these raids, though the work in rough country and bush was most exhausting and not a little perilous. Anything like a real fight was rare, for Xhosas were seldom congregated in great numbers and had only a limited supply of firearms. Still the slightest carelessness or neglect of precautions brought instant punishment at the hands of an enemy who never overlooked a blunder.

'Once, some six hundred Xhosas in a natural fortress defied a company of the 72nd, but were finally routed by the officer of the Company, who, though wounded by an assegai, managed to find a commanding position from which he fired at every head he could see as fast as his men could hand him their muskets.

'Fatigue and hunger made this fighting arduous, often there was nothing to eat but freshly slaughtered beef. "Killing and eating two hours after dark is rather a bore", wrote Harry Smith. Once in a five day raid he never shaved nor combed his hair, and only changed his pantaloons because his first pair had been torn off his legs in the bush. On the last morning he, the Commander-in-Chief, was thankful to eat a half-gnawed bone of goat, handed to him out of a filthy haversack by one of his escort. Then after a successful day, sleep at night was impossible from the lowing of thousands of captured cattle.

'To be twenty four hours in the saddle was nothing extraordinary; and once a party of men under Harry Smith's personal command traversed 218 miles of rugged mountainous country in seven and a half days.

'So swift were his movements, and so terrible his name, that the Xhosas latterly fled at the very sound of his approach. Yet they were brave men, who never missed a tactical fault and could plan and execute audacious counter offensives. More than once patrols, careless, despite stringent orders, paid for their neglect with heavy losses. But before the boundless energy and enterprise of Harry Smith the finest efforts of the Xhosa were bound to fail. By the end of May 1835, the Xhosa had been thoroughly suppressed.'

In India the requirement for mounted infantry was less due to the large presence of both locally raised and regular army cavalry regiments. This was a situation not experienced in the colonies, which, without the presence of mounted men for outpost work and quick movements, exposed any expedition to serious risks. However, one of the first occasions on which camels were used to mount infantry was in the Indian Mutiny.

On 11 November 1858 the 2nd Brigade of the Rajpootana F.F. crossed the Nerbudda and remained halted until the 14th, then marched to Charwah. Here Brigadier Parke was ordered to assume command of a column composed of light and irregular cavalry with 100 men of the 72nd mounted on riding camels, and this column was to pursue the rebels who had turned northwest and were making for Chiculda.

The detachment of the regiment was composed of the Light and No. 4 Companies all under Lieutenant Vesey. The column, under Brigadier Parke, commenced operations and after marching 250 miles in nine days, caught and attacked the enemy under the command of Rao Sahib and Tantea Topee, at Chhota Udaipur where they were dispersed with heavy loss. However, although they were scattered in all directions, they managed to reform again after some days and marched to Soloombur. Closely followed, they presently left the jungles and entered open country near Pertalghur and were met by a small force from Neemuch under Major Rocke of the 72nd, eventually escaping under cover of darkness, still hunted by Lieutenant Vesey's column, who kept a parallel course to the westward. The chase continued for several months and skirmishes took place with various columns until, in March 1859, Brigadier Parke was ordered to Jalra Patun to watch the country between Kotah and Sohait. In April, Tantea Topee was captured owing to the treachery of Nuan Sing, another rebel chief, and executed at Sepree. The rebel forces were so scattered and so depleted by casualties from being kept continually on the run that it was possible to withdraw most of the European troops to their own quarters.

Lieutenant Vesey recorded that not only did the men of the 72nd wear the feather bonnet throughout the campaign - using the feather tails to protect them from the sun - but that many of them carried their tame parrots in the hollow of the bonnet, so that it was a common sight to see the birds' head poking out from among the feathers.

Returning to South Africa we next hear of mounted infantry when, in June 1875, General Cunynghame, then Commander-in-Chief in South Africa, set about organising some mounted infantry comprising two sergeants and 38 men under Lieutenant Carrington, the Battalion Instructor of Musketry (24th Foot) being selected for this duty[5]. Most of the men could ride already and within a very short time the detachment completely confounded the critics who had declared that infantrymen could never be turned into cavalry by being proficient enough to act as advanced guard on the march from Hopetown to the Diamond Fields. The arrival of the troops at Kimberley proved sufficient to overawe a disaffected element and to enable the civil authorities to arrest the ringleaders and assert themselves. From 2 July until the middle of October 1875, the detachment remained in camp at Barkly West. When the situation became satisfactory it was ordered back to Cape Town. At Port Elizabeth, however, the mounted infantry troop were disbanded, but had proved their value and before eighteen months had elapsed, were reconstituted once again under Lieutenant Carrington, and sent to the Transvaal where they quelled a native rebellion in Griqualand West.

Meanwhile, the Trenskei natives were in a very disturbed condition, particularly in Gcalekaland, where the leader Sarili 'Kreli' kaHintsa, chief of the Gcaleka and paramount chief of the Xhosa nation from 1835 to 1892, was an old opponent of the British. It was doubtful how the Ngqika (known to the Europeans as Gaikas) and other tribes in Queen Adelaide's Provence would behave if hostilities broke out. It thus became essential to reinforce the troops on the eastern frontier. On 3 August 3 1877, the 24th Foot was sent up-country to King William's town; however, it was some time before an outbreak of hostilities occurred. Colonel Glyn, who had been appointed to command the Eastern Frontier Military District, was present with Sir Bartle Frere, then Governor and High Commissioner, at an interview with several leading chiefs but Kreli's refusal to meet the governor was a bad sign and preparation for hostilities was pushed on. Another mounted infantry troop was raised under

---

[5] Account taken from South Wales Borderers Regimental History compiled by C.T. Atkinson.

Lieutenant Clements, some of which were present at a sharp encounter with the Galekas early in October before large-scale hostilities commenced. These, however, did so the following December 1877 and on 13 January 1878, Colonel Glyn marched against a strong gathering of Xhosas between Ibeka and the River Kei and promptly attacked them. The enemy were posted on undulating ground with deep ravines on either side and, though on the right there was open ground, on the left rocks and scrub gave them good cover. Effective work was done by the rocket party of the 24th, whose fire drove the Xhosa to counter-attack, which they did with such vigour that the 88th were hard pressed. However, Lieutenant Clements brought the mounted infantry to their help in the nick of time and a decisive blow was given by Captain Upcher's detachment. This pushed forward against the enemy's flank and, after covering 400yds at the double, delivered a well-timed and executed bayonet charge, which proved most effective, completely routing the Xhosas. The attack had not started until 4pm and when the enemy retreated it was too dark to pursue them effectively, though several thousand heads of cattle were captured. It is interesting to note that this is one of the first occasions on which the Martini-Henry had been used in action.

Another action on 29 January, in which the 24th Foot and their MI took part, also resulted in the dispersion of the enemy and the capture of cattle. However, a more important success was achieved when, on 7 February, about 4,000 Xhosa, mainly Kreli's men, but partly Ngqika who had broken out under old chief Sandili, attacked Quintana. Here Captain Upcher was in command of 550 officers and men, including Carrington's mounted infantry, and learning that the Xhosas were advancing, manned some well-concealed trenches and sent the latter force to draw the attackers on until within effective range. This manoeuvre was successful in that the Xhosas advanced in two large masses and, coming on with great determination, were lured within range of the guns and rockets. This did not stop them, nor were they at first checked when the rifles opened at 900yds. Indeed, in places they got within 400yds before they wavered and broke. Carrington's horsemen now charged out to pursue, but about this time a fresh attack developed against the British right, which was repulsed by G Company. Another successful counterattack by F Company drove back a large body who had approached the right front and, about 8:30am, reinforcements arrived from Tutura, midway between Quintana and Ibeka, where Colonel Glyn had posted them. With their arrival the action was soon over and by 10:00am the routed Xhosas were in full retreat with over 250 having been killed and many more wounded. General Cunynhame described the action as 'a turning point in the war' and 'the most disastrous to the enemy of any yet fought'.

Following the conclusion of the Ninth Cape Frontier War and the annexation of the remaining Xhosa territory, the High Commissioner, Sir Henry Bartle, turned his attention to the border disputes between the Zulu leader, Chetshwayo, and the Boers of the Transvaal. Despite a commission finding in favour of the Zulus, Sir Henry, without authority from London, added an ultimatum to the Boundary Commission's findings. This was ignored by Chetshwayo and the British invasion of Zululand began on 11 January 1879 by 4,500 troops crossing the Buffalo River at Rorke's Drift, under the command of Lord Chelmsford. In the absence of any Imperial cavalry units in South Africa at this time, Chelmsford relied on the two squadrons of mounted infantry raised from the recently concluded Ninth Cape Frontier War for reconnaissance, patrolling lines of communications and vedette duties. These numbered about 300 volunteers from the 3rd, 13th, 24th, 80th and 90th Regiments. Exact numbers of MI troops are difficult to determine due to no centralisation of records and mounted infantry would be incorporated and recorded in the muster rolls and pay lists of the nearest line regiment to them.

Colonel Redver Buller commanded the mounted infantry units of the Northern column (which was under the command of Colonel Sir Evelyn Wood), and he was subsequently awarded the Victoria Cross 'For his gallant conduct at the retreat at Inhlobana, on the 28th March, 1879, in having assisted, whilst hotly pursued by Zulus, in rescuing Captain C. D'Arcy, of the Frontier Light Horse, who was retiring on foot, and carrying him on his horse until he overtook the rear guard. Also, for having on the same date and under the same circumstances, conveyed Lieutenant C. Everitt, of the Frontier Light Horse, whose horse had been killed under him, to a place of safety. Later on, Colonel Buller, in the same manner, saved a trooper of the Frontier Light Horse, whose horse was completely exhausted, and who otherwise would have been killed by the Zulus, who were within 80yds of him.'

The MI were involved in most of the major victories and defeats of the campaign, winning two VCs - Private Wassell 80th Regiment at Isandlwana and

Lieutenant E.S. Browne of the 24th at Kambula. Private J Power of the 24th was awarded the DCM for deeds at Gobatse Heights, Kambula and Ulundi.

Just prior to the First Anglo Boer War, Edwin Alderson, who became well known in the army as a trainer and leader of mounted infantry, was detached from the 97th Foot to the mounted infantry depot in South Africa. He describes the painful transformation of converting infantry to mounted infantry. 'Here again local supplies had to be relied on; all that the Ordnance could contribute were Calvary sabres and spurs, both an encumbrance when movement was perforce on foot. In addition, they had to carry their rifles and bayonets. Saddlery was bought from the local stores. However, the best part of the outfit, after the personnel, was the horses, for the Cape pony has many good qualities. In due course the sabres were discarded, and the Company took its part in the operations.'

Aldeson first saw service in the short, First Anglo Boer War of 1881, which resulted in a Boer victory over the British. This conflict highlighted the advances in development of both range and firepower of small arms that exposed the infantry to heavy losses when acting alone. A trained force of mounted infantry force handled correctly by commanders or staff with MI experience could have changed the position more favourably for the British force. As was often the case with expeditions outside of India there was little cavalry (small remnants of King's Dragoon Guards had not left Africa following their deployment in the Zulu War) available and General Colley, commanding forces in Natal, used those cavalrymen available together with volunteers from the line regiments to form mounted infantry units. With this force Colley advanced to meet the Boers on the Northern border of Natal and opting to meet them at Laing's Nek.

Laing's Nek was a defeat for the British forces partly due to the mishandling of the available mounted infantry who were under the command of Major Brownlow of the KDG. He was tasked with attacking the Boer left flank on the ridge and favouring cavalry tactics, decided on a direct, uphill, assault instead of a dismounted assault using the terrain to their advantage in line with mounted infantry tactics. This faltering attack had a knock-on effect and exposed the advancing 58th Foot to heavy Boer flanking fire, which forced them and subsequently the whole force into a general retreat. This action also exposed General Colley's lack of understanding how to efficiently use his mounted troops and the heavy losses suffered by the MI at Laing's Nek resulted in their limited involvement in the rest of the campaign and curbed the force's reconnaissance capability.

To rectify the shortage of mounted forces two cavalry regiments were despatched to Natal in February 1881. More importantly for the development of mounted infantry within the Regular Army a corps was formed from infantry regiments based in Aldershot under the command of Major Barrow. However, they arrived too late in Natal to participate in the war. This was the first example of an MI unit being formed in the UK for service abroad.

Burma had been the scene of three wars with Great Britain before she was finally annexed to the Crown in 1885. At the conclusion of the Second Anglo-Burmese war in 1852 Britain had annexed territory over what became known as Lower Burma and exercised suzerainty over the rest of Burma. By 1885 the truculent and aggressive attitude of the Burmese Court and King Theebaw's continuous intrigues with France resulted in an ultimatum being sent, demanding an apology for past insults and the reinstatement of the British resident at Mandalay, who had been forced to withdraw. This ultimatum was curtly rejected and a British infantry force with a naval contingent consisting of 10,000 men and 77 guns, was sent in November 1885. The Burmese Army was defeated, King Theebaw deposed without any great difficulty and Burma was annexed to the Crown. The pacification of the country proved a very much longer business.

There were no hereditary chiefs in Burma who could have been invested in local authority and with the native civil administration gone the whole country rapidly became infested with Dacoit bands. In the end it took four years to effectively restore order. It was soon found that mounted infantry moving rapidly through the country were the best means of dealing with the Dacoit bands, leaving the infantry to garrison fixed points throughout the country. A force of mounted infantry was consequently raised under Major Penn Symons of the South Wales Borderers to carry out this task. The force served continuously during the years 1886 to 1889 in a difficult country, the size of France, and contributed as much as anything else to its eventual pacification.

The Dacoit, like the Boers, were mostly mounted and were, in addition, particularly clever at getting away into jungles, deep ravines and even in high trees with thick foliage. Casualties were not severe, the sphere of operations was remote from India and at that time almost unheard of in England and, in consequence, the work of the mounted infantry never received the recognition it deserved. After the cessation of hostilities, however, mounted infantry continued to play their part in the army of occupation and twelve years later were to send one of the first contingents to the South African War.

Among the first British reinforcements to arrive was that of the 2nd Battalion Royal Munster Fusiliers and Brigadier General R.C. Boyle CB CMG, serving with the Royal Munster Fusiliers at the time, provided the following account: 'On arrival in Upper Burma in October 1886 the 2nd Battalion Royal Munster Fusiliers were immediately ordered to furnish a detachment of about 150 men. The Irishmen took to riding like ducks to water and soon settled down to handling the sturdy little Burmese ponies averaging from 13 hands to 13.2 hands. They were good weight carriers and ideal for mounted infantry work. They had, however, terribly hard mouths and a good deal of amusement was caused by the efforts of inexperienced horsemen to hold them in. The saddles, at first, were very rough and the ponies' backs got easily chafed and sore backs were a constant source of trouble.

'The detachment was soon split up into parties of 30 to 40 men under an officer, who generally accompanied a mixed column of Infantry and guns. These small parties were constantly in action against Bands of Dacoit; but casualties were few. Occasionally, however, young officers on their own took on greater odds then they were capable of carrying through with fatal results. On May 5th, 1888, Lieutenant R G Williamson, Royal Munster Fusiliers with a party of about 20 men and accompanied by Mr O'Dowda, a police officer, was on patrol duty from Mandalay. He and his mounted infantry attacked a strong Dacoit position and the whole party had been killed outright. News filtered into Headquarters that a disaster had occurred, and a strong search party was sent out to ascertain their fate. Confident in their recent victory, the Dacoits were awaiting them and suffered heavily in the ensuing encounter, two Burmese chiefs being among the killed. On June 25th, 1888, Lieutenant O.B. Minogue 2nd Royal Munster Fusiliers, with his party of MI came on the body of Dacoits near the village of Sepagyin led by a famous "Boh" (leader of Dacoits) named Shereyan for whose capture a reward of Rs 1,000 had been offered by the civil authorities. The Dacoits were immediately attacked and dispersed; ten Dacoits were killed including the Boh Shereyan. Lieutenant

Williamson's sword was found on the body of the dead Boh. The reward of Rs 1,000 was paid to Lieutenant Minogue and the men of his detachment. Sergeant Farrell, Royal Munster Fusiliers was severely wounded.'

Lieutenant Colonel A.J. Richardson, East Yorkshire Regiment provided the following cheery account, written in 1936, which brings a more personal touch to the story of the Burma mounted infantry.

'Some reminiscences of MI in Upper Burma in 1885-87

'There was only trivial opposition to the flotilla that carried the force to Mandalay in 1885. The Burman was, virtually, unarmed. But when King Theebaw had been deported the Burman's respect for law and order vanished and he took to a gay life of anarchy. The old order had gone and the new had not been established. It was mounted infantry that established it. In 1887 I went to the Staff Office to bid farewell to its head, preparatory to quitting Burma for India, Aden (where the battalion to which I belonged was to be stationed) and Home. Said the staff officer, "See the fellow outside there? He has just come in to surrender. He told us that three days ago he and his gang were lying in the elephant grass alongside a Choung (stream). He had some thirty men, all with guns, with him. I suppose it was you, but a party of a dozen MI came riding along within 10yrds. He wanted to shoot, but his men said, 'Don't. We are afraid to disturb hornets.'" And so we established peace that for more than thirty years was unbroken.

'In a country the size of France but nearly all wooded and with no metalled roads the infantry soldier was quite incapable of dealing with the raiding and looting gangs that opportunity bred. My first experience of one such was an order, received at night, to take out an infantry company from Mandalay, recently occupied, to deal with a big posse of Dacoits, reported beyond Amarapura. A Burmese guide was provided I put two stalwart men to take care of him and was en route in record time. Amarapura is only five miles or so from Mandalay. We were halfway there when a couple of galloping men met us. They were the head man and one of his friends, fleeing from the village to inform us that the Dacoits were, at the moment, in possession and looting with vigour. It was the winter of 1885-86 and dark. The Dacoits fired at us as, without responding, we tramped by three parallel streets right through the village, noted then for its beautiful silk industry (fortunately I had been in it more than once), and the Dacoits fled. Their old muzzle-loaders flashed at us from the fields outside, and there we did reply. When daylight came, we scoured the neighbourhood. We found a corpse or two, and I picked up a wounded man in Shan costume. He had three bullet wounds and friendlies, who had found him before I did, had slashed him well with their swords. But he recovered. I heard, when he did, that he was to be shot as a rebel. I put in special request that he might be handed over to me as a servant, as I wanted to learn Shan and was already tackling Burmese, or liberated. His account of the power of infantry armament might have a salutary effect on his compatriots. Perhaps it had. Liberated, he took a beeline for the limestone hills of the Tsawbwas.

'It was, I think, a young officer of the native Madras Regiment who originated mounted infantry in Burma. He was dispatched into the jungle to hold some river post or to patrol telegraph wires. The Burman cut these down, not from any spirit of mischief, but only to provide himself with slugs for his ancient musket. The young officer, one of those horse enthusiasts who will deny himself every luxury and even necessities to maintain a stable, usually of one, fortunately, had with his detachment a few transport ponies. His post established, he put blankets on his pack saddles, seated his Ramaswamies on those, and captured, or chased out of the district, every undesirable character. The VC came to him before he returned to the racecourse, the snipe and barrack routine.

'Late in 1886 the efficacy of mounted infantry was recognised by the Army headquarters in Upper Burma. Companies were formed, but with very crude equipment. The Burman pony is small. Saddles were not available. Dreadful substitutes were invented. Things without "trees", made of flock stuffed into cotton covers, things that sore-backed the poor ponies in a few hours, because on them, his feet in his stirrups, set a well-fed soldier, with no experience of horses, but carrying a heavy Martini or Snider rifle and a supply of ammunition the bullets of which, weighed more than a whole modern cartridge.

'Down near the frontier between Upper and Lower Burma a company had been formed. Its establishment were twenty-five British soldiers and fifty native (Baluchis - usually horsemen). The officer commanding the company knew no Urdu, the *lingua franca* of the Indian Army. There was friction between British and Baluchi. The officer's ill health let me step into his shoes. And then began the time of my life. I was alone, commanding seventy-five men, seventy-five ponies, no veterinary assistance procurable, away out in the jungle, chasing elusive "bo's" (leaders) for days at a time, with three quarters of my unit left at our station in the care of one of Her Majesty's privates, walking or leading our ponies through the night with short bursts of riding, to lie down when within reach of our objective, reins hooked to hand or foot, and snatch a few minutes sleep before the stars begin to fade down on the horizon, and we mounted to surround a *kyaung* (monastery-school) or village before its inhabitants awoke with the dawn. Only twice did we kill, but efficacious? Don't disturb the hornets.

'My infantry brigadier had tried to get me a subaltern but I was so junior that it was quite a long time before he did. Shortly afterwards things began to hum. Proper saddles came from Cawnpore, India, and the order went forth that my company was to be doubled and divided. I handed over the Baluchis to an officer of that regiment, received 50 privates from the British battalion at the station, ponies and equipment. Three days sufficed to teach a man to ride well enough to come out "bo" hunting and, although escorts were held necessary for all individuals or transport coming to or going from our station, I could boast that by night or day, with any one of my men, I could ride in safety anywhere within 100 miles of it. A knowledge of Burmese, sketchy but adequate enabled me to converse with natives without the intervention of an interpreter. Thus, I could get information of trouble ahead. The natives would divulge nothing if an interpreter was present, having no assurance that the Dacoit sympathisers would not hear of the disclosure, with painful results to the informant. The Burmese Buddhist takes no life, normally. When he throws his "Yan mingalan" overboard he can be as cruel as a Torquemada.

'In the autumn of 1887, I left – to hear some months later, on a transport in Bombay harbour, that anthrax had killed about 80% of all animals at the station. We had an anthrax scare in my time. Transport animals had died of the disease. I had wasted no time. Within three hours my company was on its way to a beautiful spot, with good grazing and a lovely little river, twenty miles away. It was there that I had left it. My successor, a captain in the regiment supplying the men, had brought the unit back to civilisation. The men liked it; the ponies died.

'Penn Symons[6], he had hunted hounds at Octacamund when I was up there on leave and he died of wounds at Talana, one of the early fights in the Boer War. He was commandant of all the MI units in Upper Burma. He loved a horse, whether it was a 12.3 Burma pony, a 17-hand hunter or one of the team that he drove so skilfully up and down that zig-zag mountain road of the Nilgheries. But that generation has passed and gone. The horse is obsolete, and the aeroplane dominates the art of war.'

[6] Later Lieutenant-General Sir William Penn Symons KCB (17 July 1843 – 23 October 1899).

Chapter 3

Egyptian campaigns of 1882 and 1884

The Egyptian campaigns gave a considerable fillip to the reputation of the mounted infantry and advanced its progress from improvisation to a permanent system of training for use, as required, on service.

On 17 July 1882, the troops landed and immediately a body of mounted infantry – two officers and thirty men – was formed under the orders of Captain Hutton[7] (who was to play a major role in the future development of the mounted infantry). The Khedive supplied his private horses for the purpose. They were an excellent breed and proved invaluable. The men were taken by careful selection from the battalions landed, i.e. the Stafford's and King's Royal Rifles. Both these battalions had served in South Africa, and the men selected had been among those employed as mounted riflemen; those of the King's Royal Rifles having belonged to Captain Hutton's own corps, which no doubt, accounted for their great efficiency under somewhat similar conditions.

On 18 July a party of MI patrolled 10 miles towards the enemy, and from this time onward daily rendered most valuable service in that capacity. On 22 July the railway in front of Mallaha Junction on the enemy's side was destroyed by a force under Captain Hutton, while companies of the rifles and four naval guns formed a support upon the high ground above Ramleh under Colonel Ashburnham. The work was carried out without loss.

By the middle of August, the moment had arrived for putting into execution the first part of the programme of the campaign, which was as follows:

a) To obtain possession of the Maritime canal in order to ensure the safe passage of the expedition.

---

[7] Later **Lieutenant-General Sir Edward Thomas Henry Hutton, KCB, KCMG, DL, FRGS** (6 December 1848 – 4 August 1923). In 1888 he was appointed the first Commandant of the School of Instruction for mounted infantry, Aldershot.

b) To move to Ismailia, in such successive order as the conditions permitted the force to be employed.
c) To keep the enemy by every artifice under the impression that the attack would be made from other points than Ismailia until a force had arrived there large enough to ensure its safe possession.
d) To seize and secure the line of railway and the Sweet Water canal to within striking distance of Tel-el-Kebir.
e) After these lines of communication had been safeguarded, to put in order such parts of railway or canal as the enemy had obstructed or damaged, and to place upon them such rolling stock and storeboats as would render them available for the supply of the army in the field.
f) To accumulate at an advanced post, under an adequate guard, stores and ammunition sufficient to make the army, when it moved forward, independent, except for the purposes of distribution, of all other modes of transport.
g) To engage the enemy in general action when, and not before, the accumulation of stores permitted the concentration of the army for the delivery of an effective and, if possible, a final blow.
h) To follow up this action with the greatest rapidity, in order for political and military reasons, to secure possession of Cairo and to seize the points of concentration to prevent the numerous Egyptian forces at Kafr ed-Dariar, Cairo, Damietta, Rosetta and Es-Salihyeh uniting into a formidable force.

Consequently, on 23 August orders were issued as follows:

> 'The following movements will take place: The three squadrons of Household Cavalry; two guns of N Battery, A Brigade, Royal Horse Artillery; the detachment of the Nineteenth Hussars and the mounted infantry will march independently tomorrow morning at 4am for Nefisha and will place themselves under the orders of Major General Graham.
>
> 'As soon as Major General Graham has been joined by the above troops, he will proceed to Magfar and take up position there.'

Though the desert on the eastern side of the Delta is generally hard, excellent marching ground, the portion of it which lies immediately between Ismailia and Magfar is relatively soft. Hence this part of the route was particularly trying for all ranks and the artillery were compelled to follow the railway embankment by way of Nefisha. Sir Garnet Wolseley, with a certain portion of his staff, together with General Willis, accompanied the cavalry to Magfar, in advance of the infantry.

From prisoners captured on the march, it was ascertained that the Egyptians had constructed a second dam at Tel-el-Maskhuta, and that the infantry were there in force behind entrenchments. The apparent determination of the enemy to hold on to Tel-el-Maskhuta offered a new prospect. For about half an hour General Wolseley carefully watched the enemy lines. He saw trains arriving at the station and bodies of Egyptians beyond the line of entrenchment, which convinced him that a serious stand was contemplated. Despite having few troops available and the prospect of a long march he realised that the very smallness of the force was likely to tempt the Egyptian commander to stay where he was, thus giving time for the British reinforcements to come up from Ismailia. If he delayed sufficiently, it might be possible to inflict upon this outlying body a defeat which, if followed up, would enable them to secure the very line of railway and canal as far as Kassassin, which, at that time, was all important.

Whilst Sir Garnet had been examining the enemy position, the advanced troops had reached Magfar and, on making up his mind that he would now have an opportunity of fighting an action of some importance, he ordered General Willis to take command. Willis ordered that all were to lie down under cover of the scrub and neither show themselves nor fire without special instructions. The cavalry and mounted infantry were considerably to the right, about a mile from the canal, watching any attempt of the enemy to turn the right. At 9am the enemy opened on them a distant but heavy musket fire. Thereafter for nearly an hour the mounted units continued to hold the enemy's left in check on the hard, gravelly flats of the hill tops. It was reported that, as soon as the bolder Egyptian spirits attempted to advance, they were instantly picked off by the men of the mounted infantry, who, being all chosen marksmen, had dismounted, and by their effective shooting for a time, stopped all movement on that side. Owing to the encircling position of the enemy, the mounted infantry were, throughout nearly the whole day, in much closer contact with the enemy than any other portion of the force with the exception of the extreme left detachment.

Sir Garnet, with his staff, had been standing upon the top of the hillock above the position occupied by two guns. At 9:40am the enemy opened fire with four guns on this hillock and had obtained the range with such accuracy that the first shot struck the foot of the rise; the second passed a few feet over the General's head exactly in the right line and a third followed with as good an aim, breaking a horse's leg just behind the hill. However, as soon as all the horses had been led out of sight so that no conspicuous object was in view, the guns were turned on the mounted infantry, firing so accurately that they were ordered to fall back. The wide plain afforded them little shelter, but fortunately, despite brilliant laying, the shells themselves were very inefficient. The shrapnel, which seemed to be scarce, burst high in the air as they were fitted with time fuses, while the shells, fitted with percussion fuses, were intended to burst on touching the ground but, owing to the sandy soil, they buried themselves. Unless they actually struck a man or horse, were almost harmless. Consequently, British losses were light. Nevertheless, lack of water caused a great deal of suffering, since there was no shade of any kind, and eventually General Willis sent his own aide-de-camp across with the precious fluid, together with reserve ammunition.

On the following day, General Graham's brigade moved along the canal in attack formation, since the enemy were gradually retiring. Their route lay along the canal and as soon as they came within sight of the enemy's position it became evident that the Egyptians had altogether abandoned the works. Therefore, the infantry advanced towards Tel-el-Maskhuta , while Sir Garnet sent orders to the cavalry and Horse Artillery to work well round the enemy's left and cut off his retreat, if possible. The cavalry horses were scarcely fit for such severe exertion so recently after landing from shipboard. However, the small Egyptian animals on which the infantry were mounted, well-bred and accustomed to the country as they were, enabled the little force to cover the ground with greater ease. They therefore manoeuvred accordingly, but no formed body of the enemy was seen until the heights over Mahsama were reached. Then Sir Drury Lowe found himself opposed by a large force of infantry, with guns in position, and eight or ten squadrons of cavalry. The artillery had meanwhile been reinforced by the two guns from A Battery, 1st Brigade, at first on right of the Guards. The shrapnel proved extremely effective, as did the fire from the mounted infantry, employed strictly in their proper function of galloping up into position, dismounting and bringing close and well-aimed infantry fire to bear. As a result of these tactics, the enemy successively fell back until, at last, the camp at Mahsama was captured. Seven Krupp guns, a large number of Remington rifles, a great quantity of ammunition and stores, together with a supply train of 75 wagons, fell in British possession.

After reaching Kassassin, where some brisk fighting was recorded, reinforcements came up and the mounted infantry were attached to the Indian cavalry brigade under General Wilkinson. A general engagement followed at Tel-el-Kebir, in which the mounted infantry were heavily engaged, but it was after the big battle that they once more proved their use as a mobile force. Under General Wilkinson they marched to the north side of the canal in advance of the main body to Aabasa Lock, where they crossed, moving along the western side upon Belbeis, which was seized on 13 September.

Captain Watson R.E., on taking over the telegraph office, found messages from Arabi Pasha to the officer in command at Es-Salihiyeh ordering the Egyptians there stationed to be sent at once to Mansura by train and also repeated requests for information as to the movements of the English. This rapid pursuit on the part of the cavalry already disconcerted the attempt of the Egyptian commander to gather fresh forces for action. Another important advantage was gained by the opening of the lock sluices so as to send a fresh supply of water down to Tel-el-Kebir.

Next day the advance was resumed and, as it progressed, the local inhabitants on the other bank of the canal crowded down to the water's edge, receiving with acclamations of pleasure copies of a proclamation, previously prepared in Arabic, in which Sir Garnet informed them that the war was over and that peace now reigned in Egypt. Upon their arrival in Cairo, Captain Watson was despatched to the citadel in command of a party consisting of two squadrons of the 4th Dragoon Guards (five officers and eighty-four men) and mounted infantry (four officers and fifty-four men). At the main gate the men were halted and closed up. Captain Watson rode on to find that nothing had been done towards evacuating the citadel, which was still held by the whole garrison, at least 5,000 strong. He therefore rode into the stronghold and arranged with the governor, Ali Bey Yousif, that he should parade the garrison and march them out. At long last the Egyptians passed out by the Bab-el-Ayab gate, watched by an officer and a few men – the whole arrangement being treated, as far as possible, as a simple relief of the Egyptian garrison by English troops.

The keys of the citadel were handed over to Captain Watson as, regiment by regiment, the men marched past, out into the city to the Kas ren-Nil barracks which had been assigned to them for the night and where, the following day, they were to be quietly disarmed. As the last troops were leaving, the guards were relieved by the tiny party of English soldiers and Captain Lawrence of the MI was left in charge of the citadel.

The Egyptian campaign of 1882 had been brought to an end by Lord Wolseley's success at Tel-el-Kebir and the subsequent occupation of Cairo. Egypt itself was quiet but the fires of revolt were smouldering in the Sudan. A Mahdi had arisen and he annihilated the Egyptian forces sent against him, capturing a great store of guns, rifles, ammunition and supplies. The garrison at Khartoum under General Gordon was beleaguered and though his orders, in taking command in 1884, had been to withdraw the garrison of Sudanese and Egyptian troops, he had decided to stay as the withdrawal would have entailed leaving behind many of whom the Dervishes would have made short work.

At the beginning of 1884, a force from Egypt and India, which included mounted infantry, was landed at Suakim. Several actions were fought at El Teb and Tamaai, but nothing was really gained and the force reembarked at Suakim. The Eastern Sudan remained in the hands of the Mahdi's Lieutenant, Osman Digna, except for the actual coastal towns which the British held.

Gordon's position was now precarious and the British government was loath to further commit itself. There was much heckling and plain speaking in the House since public opinion was strongly in favour of Gordon's immediate relief. It was at last decided to send an expedition to the Sudan to assist in the withdrawal of General Gordon and his garrison from Khartoum.

In the spring of 1884 Lord Wolseley entrusted Major Hutton with the task of drawing up a practical scheme for developing the force of mounted infantry in Egypt and, since Captain Humphrey then in command there had been killed at Cairo in a steeple chase, he was sent out to put his promises into effect.

In view of the country to be crossed to reach their objective they were to be mounted on camels and were styled the Mounted Infantry Camel Regiment (MICR). The Khartoum Expedition planned for the following autumn was now well under way using camels with a very limited number of Arab horses as mounts.

To increase the troops available in Egypt, two companies of mounted infantry were sent out from England to join the two companies already there. Each company consisted of four separate sections of one officer and twenty-eight other ranks from different infantry battalions, with a captain to command each company. With the usual complement of officers and other ranks as regimental staff, the strength of the regiment was some 25 officers and 400 other ranks. Of the two companies formed in England in August 1884, one consisted of two divisions each from the King's Royal Rifle Corps and the Rifle Brigade. The second was from the Somerset Light Infantry, the Royal Scots Fusilier, The Queen's Royal West Kent Regiment and the Connaught Rangers. The company was commanded by Captain Pigott, formerly of the King's Royal Rifles but then serving in the 20th Hussars. He had seen service in the 1881 Boer War, in Egypt and in the Sudan.

Hitherto it was supposed that white men could not manage camels without the aid of native drivers. Hutton, with the aid of Colonel Robert Tudway, thirty men of his Essex Division and a similar number of camels, was able to disprove this theory within four days. He was then able to demonstrate to General Sir Frederick Stephenson a number of mounted infantry movements on camel back which speedily convinced the authorities that here was the answer to the cavalry problem. However, it was not all plain sailing as Lieutenant Marling of the Rifles MI later wrote about a mounted parade held by the Camel Corps on 27 October 27: '[We] did fours, right and left and wheeling, and only about six men fell off. Then Curly Hutton sounded the trot, and in two minutes the air was thick with Tommies flying about at every angle. Twenty-three camels got loose and went off with their tails in the air toward the setting sun, and we never got back five of them. Curly Hutton came off on his head.'

A Camel Corps was thus formed with Major Hutton in command. Unfortunately, the severe task of training this unit in heat, together with two bad attacks of sun syncope, resulted in his being sent home a very sick man so that Major Grant actually took the corps into action.

The following abridged version of Lieutenant-Colonel M.J. Duggan OBE's diary records his experiences serving as a Quarter Master Sergeant in the MI during the campaign. The full account was first published in the Journal of the Royal Scots Fusiliers circa 1934.

'Our mounts were a good class riding camel, saddlery and equipment carefully thought out and sent from Aden. The saddle weighed 64lb., was well padded and fitted with a pair of large canvas saddle-bags, a leather water-bottle to hold about half a gallon and a mussick to carry about four gallons. The mussicks, when not in use on the march, were kept full of water and hung on a tripod in order to prevent leakage from dryness of the skin.

'Our commanding officer and adjutant left on their camels for Dongola on the 28th inst. [October 1884] but returned the same evening as Major Hutton was suffering from sunstroke. He was invalided down country and did not re-join during the campaign. He was a wonderful organiser. I served under him again for some years in the training of mounted infantry after the war.

'D Company left for Dongola on 31 October, leaving behind Captain Phipps, 7th Hussars, who was our quartermaster, myself, and about a dozen men to bring on the remainder of our stores and equipment. Lord Wolseley, a staff officer, and his Egyptian ADC left for Dongola on the same afternoon. We left with our camels and stores for Dongola on 4 November, and reached our destination on the 17th inst. During the march we experienced very severe sandstorms. The march was trying at times and particularly so when it was necessary to leave the river and strike out to the desert when going around hills. We passed Lord Wolseley and staff on his return journey from Dongola on 13 November. He must have been doing well over forty miles a day, and at the same time carrying out his inspections. We crossed with our camels from east to the west bank on native boats and joined our headquarters at Dongola. The small supply posts between Wady Halfa and Dongola were held by small detachments of the 1st Royal Sussex. The headquarters of the Desert column was at Dongola, with Major General Sir Herbert Stewart in command, Major Wardrop, A.A.G. and Captain Lord Arlie, Brigade Major.

'Major Gough, of the 10th Hussars, who had been our second in command, was appointed to Command, Major Barrow was appointed second in command and Captain Phipps, who had been our quartermaster so far, was appointed to the vacant majority. I was very sorry to lose him, as he was always ready to help me in any difficulty. We left Dongola on 25 November for Handuk, arrived there on the 27th and remained until 10 December. We managed to start a native bakery here and supplied our men with fresh bread daily.

'The Guards Camel Regiment joined us during our stay at Handuk; neither their camels or equipment were equal to ours and had evidently been hurriedly got together. Between Dongola and Handuk I pushed on by myself on the second day in order to obtain green forage for our camels and after having succeeded, was preparing to distribute it along the camel lines when Sir Herbert Stewart, who was in command of the Desert column, Major Wardrop, and Lord Airlie arrived on the scene, having landed from a small launch, and after talking to me and hearing what I was about to do, insisted on helping me. They put down the green forage for A, B, and C companies and I put down that for D Company.

'The Nile boats which we had seen start from Gamaai on 5 November commenced to pass up the river at Handuk about 10 December, but not in the order they had started. Some got through over the cataracts fairly well, but others were driven back by the strength of the current and often reached the cataracts at their most difficult points. This often meant the loss of lives and boats. Canadian boatmen, Kroomen and Dongolese assisted in the passage of the cataracts by the boats. When we left Wady Halfa the troops of the expedition were divided as follows: One battalion at Alexandria, one cavalry and four battalions at Cairo, one battalion at Assuan, one battalion at Koroska, one battalion at Wady Halfa, one battalion and one mounted infantry regiment between Wady Halfa and Dongola. Most of these troops were now moving up the river for concentration at Korti, where Lord Wolseley had his headquarters. Only small forces were left in Lower Egypt.

'We left Handuk on 11 December and arrived at Debbah at 12 noon on 13 December. There had been a fight between the Mahdi's men and the Bashi-Bazooks (Turkish irregulars) who were holding the fort at Debbah. The Debbah garrison had been victorious. The dead were still lying about and no attempt had been made to bury them. The native force, under the command of Captain Kitchener of the Royal Engineers, who was employed on intelligence duties, was drawn up as we passed the fort. He had been the only European in this part of the country for some months and was, I am certain, glad to see our advanced troops appear at last. I may mention that this was Lord Kitchener of Khartoum, who afterwards reconquered the Sudan in 1898.

'We arrived at Korti, our advanced base, on 15 December. The weather, so far, had been very kind – no rain, but cold at night. We were ahead of our supplies and had to do the best we could on preserved meat, biscuits and what fresh vegetables could be procured. Lord Wolseley and his staff arrived on 16 December. We heard rumours that Khartoum was hard pressed and that our force should push on as soon as possible.

'Between our arrival at Dongola and the 29 December the boats carrying the River column continued to turn up well and the concentration was being pushed on. On Christmas Day the troops were given extra rations of date, flour and rum. At night we had a grand concert. The Heavy Camel Regiment joined us at Dongola. Very big men, but much too big for the short rations that we had to live on. Our men, the Mounted Infantry Camel Regiment, were medium height and weight and just the right stamp for this work. On 30 December we received sudden orders to march and left at 3pm together with the Heavy and Guards Camel Regiments, making due west and into the desert. Our guides were pressed men from the district and not to be trusted. Lord Wolseley watched us pass about 2 miles from Korti. At this date we had only evacuated one man of the Royal Scots Fusiliers Division for sickness. The march continued until 8am, 31 December, when we rested until 1pm. Continued the march until 3am, 1 January 1885 when we rested until 6am and then marched on and occupied Gakdul Wells at daybreak on 2 January. Our camels marched the ninety-six miles in sixty hours with only eight hours rest. This was good going for camels whose ordinary pace is two miles an hour. We found no sign of the enemy at the wells. On the forced march to Gakdul our greatest drawback was lack of water on the way. What wells we passed were practically dry and of no assistance. The camels could not be watered, and our own supply meant using up what we had in the waterskins. This was meant to be conserved in case of there being no water a Gakdul.

'An Arab bandit was captured during the midday halt on the 31st. He had his wife with him and used a donkey for transport. He could only hobble about as his feet had been so unmercifully flayed at some time they were cut to the bone. He had been engaged in intercepting messengers from both sides and relieving them of all their valuables and papers. He told us his wife would be all right if left, so we took him along with us. He was attached to the MI for rations and retained his donkey for transport.

'On the morning of 1 January 1885, we intercepted a date convoy making for the Mahdi's force and brought it to Gakdul. We were able to water all our camels at the lower well at Gakdul and the upper wells in the surrounding hills were used for drinking. They contained rainwater, which was very brackish but, nonetheless, welcome. The rains only happen at intervals of five to eight years and I was told the last rains had been six years before. One can well imagine what the water was like. Having seized the wells, the Guards Camel Regiment were left at Gakdul to prepare it for defence and improve the water supply. Their camels and saddlery were handed over to the mounted infantry to take back to Korti while Captain Kitchener remained at Gakdul. We left our return march at 8pm the same day and arrived at Korti early on the morning of 5 January, having marched 192 miles in five days, twenty hours and having only seventeen hours for sleep. The men were worn out on arrival and our camels had suffered intensely. Many of the camels dropped out on the return march and, as a whole, were never able to recover their strength during the subsequent marches.

'We left Korti on 8 January for Gakdul and had with us the remainder of the force for the Desert column, viz., staff, one squadron of 19th Hussars under Major French (the late Earl of Ypres), two camel regiments, one infantry regiment (mounted on Guards camels), medical and commissariat detachments, naval detachment under Lord Charles Beresford with one gun; four screw guns under Captain Norton R.A., and the usual native camel drivers for the hospital; also about a dozen war correspondents. We arrive at Gakdul on the 12th. There was very little moon during this march, so we were able to have longer rests. The Light Camel Regiment remained at Korti for convoy duty across the desert. The headquarters and six companies of the Royal Sussex, after handing over the Guards camels to this unit, remained to garrison Gakdul. Two companies of the Royal Sussex came on from Gakdul with the column. Colonel Sir Charles Wilson R.E. caught us up at Gakdul and took over the duties of chief intelligence officer.

'On 13 January the column left for Metemneh with a strength of about 1,500 all told, four screw guns and the Naval Nordenfelt. As we moved out from the wells, I remember seeing Captain Kitchener sitting on a rock and watch the column go. It must have been most disheartening to him to be left behind after all the dangerous and lonely work he had carried through for over a year. At about 3pm on the 16th our cavalry scouts (19th Hussars) reported that a large force of Dervishes was between us and the Abu Klea wells. It should be understood that our camels were only used as means of transport; we fought and acted as Infantry when in action. We were ordered to dismount, hobble down our camels and form up in square just at the entry to the wady. After forming up, Sir Herbert Stewart, our commander, issued orders to form a zereba, where our camels were, and put the position in state of defence for the night. The mounted infantry sent two divisions to occupy a high hill about 400yds on our left flank, with instructions to build a breast work and hold it until withdrawn. During the evening and that night, the enemy kept up a heavy rifle fire on our right flank. It was too late that evening to attempt any dislodgement, so the enemy's fire continued throughout the night. Many casualties occurred amongst the troops and camels.

'At daybreak on 17 January, we had breakfast and formed up in square, just in front of our zereba. All camels and stores were left in the zereba under certain details from each regiment. At 8am our screw guns opened fire on the enemy. The position was noticeable from the number of flags flying, but we failed to move them or obtain any idea of their strength or disposition. The 19th Hussars had moved out on our left flank to cover the advance. The square commenced to advance about 8:30am with the hospital cacolets and camels in the centre, and the guns (four) inside and supporting each face of square.

'The enemy still kept up a dropping fire from our right flank and as casualties occurred, the men were attended to and placed in camel cacolets. When about 300yds from the zereba, two of our men (RSF), White and Rafferty, were hit almost simultaneously. The square continued to move on. I left the square and found both men were alive but badly wounded. I succeeded in attracting the attention of those in the zereba and when sure help was coming, I re-joined the square. White died of wounds, but Rafferty recovered.

'As the square continued to advance along the high ground and towards the enemy and wells, the mounted infantry sent out two officers and about twenty men as scouts on our left flank. The square continued to advance and gradually the scouts moved towards the flags where the enemy was supposed to be. This position was about 600yds from the square. Up to this no movement was observed but as soon as the scouts reached a position overlooking the wady, the enemy rose and came direct towards the left face of square – in an echelon from the centre formation, three or four mounted Emirs, on gaily-caparisoned Arab horses, led the advance and the whole came at a very steady slow run. The scouts at once retired direct on the square with the enemy close behind, but without breaking their ranks by attempting to overtake the scouts. The scouts being tired out and carrying 200 rounds of ammunition, could only retire slowly, and this meant the enemy were able to come on until the scouts reached the square and allowed us to open fire. One of the scouts unfortunately dropped from fatigue when about 150yds from safety and was no doubt speared as the enemy passed over him. As soon as the Dervishes started their advance the square halted and came to the ready position – front rank kneeling, rear rank standing and remained so during the retirement of the scouts and until our front was clear. The fire discipline to this point was magnificent, and everyone seemed to realise that should one shot be fired before the scouts cleared the front there would be little chance of preventing independent firing. It appeared to me that fire was opened on the enemy by the left face when the leading ranks of the enemy were less than 100 yds from us. It was independent, well directed and low.

'The enemy came on until the leading ranks were shot down and formed a barrier over which those behind could not pass in time. The mounted Emirs still came on with a certain number until within about 20yds of the square, when they were shot down. The enemy's front overlapped the left face considerably and a party on the enemy's right flank succeeded in reaching the left rear face of the square. Accounts say that this part of the rear face had been moved up under orders of Colonel Burnaby to extend the left face and prolong the field of fire during the enemy's advance, and that it was at this spot that the enemy broke in. I personally did not notice this movement at the time although I was well down the left face. When the enemy was stopped by the fire of the left face they turned back and retired sullenly. In the meantime, those who had reached the square were determined to kill where possible and, although the hand-to-hand fighting was only for few minutes, our casualties were very heavy, but none of the enemy succeeded in leaving the square. Although the cease fire was given as soon as the enemy started to retire and the square was cleared, odd Dervishes continued to rise from the ground and charge the square. Major French's squadron of hussars appeared on the enemy's flank as soon as the retirement started, but as the horses had not been watered for certainly two days they were not fit to charge and, in any case, the enemy was far from demoralised.

'The square was reorganised at once and our casualties during this period were nine officers and sixty-five men killed and ninety wounded. Exclusive of this, most of the native hospital camel drivers had been killed and a number of camels carrying wounded men. The whole of the Naval Gun Detachment were killed when the gun jammed. Lord Charles Beresford, in command of the naval contingent, who was suffering from an abscess on his leg, was mounted on a donkey during the fighting. His donkey was knocked over by a camel falling and Lord Charles was under the donkey. Only for this, I think there is no doubt, but he would have been killed the same as the gun's crew and officers. The front and right faces were not attacked but, I was told afterwards, that the enemy's riflemen opened a heavy fire on the square during the charge. The enemy's strength was said to be 10,000 and their casualties must have been at least 3,000. Our fighting strength in the square would not exceed 1,000. Our commanding officer, Colonel H. Gough, was wounded early in the advance.

'As soon as we reformed and each man had received about half a glass of water (the three war correspondents who were with us also received a ration of water, viz., Pearce, Piggott and Ingram), we moved direct for the wells, cavalry covering the advance. We reached the wells of Abu Klea, at sunset, but found very little water. I think it was during the square's advance that Lieutenant Stanuell, who was commanding the Royal Scots Fusiliers Division, got sunstroke. The wells were only sand holes and it took some time to get any quantity of water through and, even then, the supply was very limited.

'As night was drawing on, we were formed up in two ranks in a square formation and told to rest. We soon went off to sleep, but in less than an hour I was sent for and told by Major Phipp, our second in command, that he was taking back a party to the zereba to bring on the camels and stores and wished me to come with him. The party was about 100 strong. The march back over the battlefield was trying in the extreme and as the party left in the zereba could not be notified of our return, we had to move with the greatest caution. We struggled on, although at times dropping into holes and ditches. When in the neighbourhood of the zereba we halted. Major Phipps moved forward and was challenged by a sentry, and at once replied. We waited for a short time and were told to advance. The garrison at the zereba had stood to all night, as they had no idea how the fighting had gone (except for the sound of heavy firing) after the square disappeared over the high ground. The men in the zereba at once made some tea and we were told we could sleep for two hours. It was just before daybreak when we reached the zereba. After our rest the camels and stores were formed up, the two divisions of mounted infantry holding our left flank were recalled and we started, together with the remainder of war correspondents who had been in the zereba, on our march back to the wells. We had a very good view of the previous day's battlefield and realised how severely the enemy had suffered in killed and wounded. We were not at the time able to do anything for the enemy wounded, though afterwards some of them managed to reach the wells and receive attention. On arrival at the wells, we received orders to march that afternoon. When going over to the commissariat to arrange supplies I heard our commander soundly rating one of the war correspondents for sending back to Korti without his sanction a native messenger with the account of the action at Abu Klea. An entrenched position was formed at the wells and our wounded attended to. The two companies of the Sussex Regiment (except a small party which was intended for Khartoum) were left as garrison.

'The Abu Klea Wells are about 18-20 miles direct east of Metemneh and the Nile, our present objective. We left for Metemneh at 3pm on 18 January and marched all that afternoon and night, expecting to reach the Nile near Metemneh before daybreak on the morning of the 19th. We passed through rough ground and trees during the night and frequently had to halt in order to close up the ranks of camels. Several times I found myself sleeping and my camel amongst strange units. This was the case with everyone, as we had been without practically any rest since the night of the 15th. As a matter of fact, only two men (one of the Heavies and one of the Guards) and a native boy servant strayed from the column. The man of the Heavies and the native boy were captured by Dervishes, but the Guardsman, who was leading two ammunition camels, succeeded in retracing his way back to Abu Klea, although chased by enemy scouts when near it. At daybreak on the 19th, we were still some distance from the Nile and by sunrise we found the enemy in scrub all round us. The

Arab bandit, who was still with us, told our interpreter that our native guides had deliberately misled us and that we should have reached the Nile very early that morning. When we halted, we were about five miles from the Nile and as the enemy soon commenced to fire from the scrub, we were ordered to form a zereba of boxes of supplies and camel saddles in case of attack.

'The men were completely worn out by now and it was arranged that while one half slept, the other half would reply to the enemy's fire. Some of the men (Royal Scots Fusiliers) under an officer of the Royal Engineers (afterwards Lieutenant General Sir Henry Lawson), were ordered to take out biscuit boxes to the entrance to the scrub and build a small breastwork for defence. Private J. Stewart (afterwards Colour Sergeant in the 1st Battalion and depot) was one of this party, and when discharged to pension I was glad to be able to help in obtaining an increase in this pension for his gallant conduct. Casualties were now rather heavy and early in the morning our Commander, Sir Herbert Stewart, who was standing alongside our division was badly hit and taken to the aid post. This was a great loss; he had the fullest confidence of all officers and men with the Desert column. In point of seniority the command now devolved on Colonel Sir Charles Wilson, R.E., but, being in charge of the Intelligence Branch, he handed over command to the next senior officer Lieutenant Colonel The Honourable A. Boscawen, of the Guards. Up to 1881 all Lieutenants of the Foot Guards had also the rank and seniority of a Captain in the Army, and Captains in the Guards were Lieutenant Colonels in the Army.

'Our position was now desperate; our number of wounded was increasing every moment; our water supply was practically finished, and it was imperative we reached the Nile. The first order received through our Brigade Major (Lord Airlie) was for the mounted infantry to fight to the Nile. We had only 300 men left. Later on, this order was cancelled, and it was decided to hold the zereba with a portion of the Heavies and other details including cavalry and guns; and, the remainder of the Heavies, the Guards and mounted infantry should fight to the Nile, establish ourselves there if possible by night and return next day with water to the zereba and bring back the wounded and stores to the Nile. During the morning, St Leger Herbert of the *Morning Post*, and Cameron, of the *Standard*, war correspondents were both killed. Our casualties up to leaving the zereba were 14 killed and 60 wounded.

'At 3 o'clock we received orders to form up on the right side of the zereba facing the Nile – there was no question about remaining in the zereba; we were without water and with luck we might succeed in forcing our way through. The screw guns were to be kept in the zereba and cover the square's advance on the river and, from this position on the high ground, could help the advance. Immediately we formed up, the enemy poured in heavy rifle fire from the scrub and caused many casualties. The square was formed with the Guards right front and right faces, mounted infantry left front and left faces, the Heavies the rear face. The four divisions composing D Company of the mounted infantry were ordered to support the corners. The Royal Scots Fusiliers were in support on the right front and right face corner. The centre of the square had the ammunition camels and about another half dozen with cacolets for wounded. The enemy's fire hastened our forward movement in the direction of the low ground and the Nile. Almost at the start Carnegie was hit in the leg but was able to crawl back to the zereba; another of our men, whose name I cannot call to mind, was also hit. The square's fighting strength was about 700. At the beginning of the advance the enemy's rifle fire caused us to halt and lie down two or three times, but nothing was gained by this, and it was decided to push on to the river. The guns were helping as well and the direction of their fire, which was over the high ground in front of us, thus gave us knowledge as to the enemy's position and we marched accordingly. Major Phipps, of the MI, was acting as guide for direction of the square about 150yds in front. The front and left faces were directed towards the enemy so as to open fire as soon as they appeared. The enemy delivered their charge at about 4:30pm and in the direction of the left and front faces of the square. Led by mounted Emirs, they came over the hill numbering probably 12,000. On this occasion we had no scouts out and Major Phipps had come in, so our front was clear as the enemy charged down the hill. The square halted, cheered and many helmets were thrown up and then, with the front-rank kneeling, the front and left faces opened a terrific fire.

'The enemy's formation was broken almost at once. The Emirs were shot down, the rear formation of the enemy turned back over the hill, and we were ordered to cease fire. I suppose the idea was to reach the river without delay, rather than wanting an enormous casualty list of the enemy. The ground was covered with the enemy dead and wounded, but we moved to our right so as to avoid this ground and marched to the river. Our guns ceased firing after the enemy charge was stopped, the enemy riflemen moved off towards Metemneh and made no further attack on either the zereba or square.

'We reached the Nile about sunset and, after being allotted our positions for the night, the men were sent down to the river for water. My batman had brought his camel mussick attached to his belt and, after obtaining a drink he filled the mussick and, I noticed, several of the men having a satisfying drink from it as I dropped off to sleep. Just as we were reaching the river I had a look around the square and noticed my greatcoat was thrown across one of the camel saddles, but as I had not put it there and Brown said he had not, I could only imagine that when starting from the zereba someone had noticed the coat and thought it would be useful – so it was: it was the only cover, except for camel nets, the wounded had that night.

'The discipline and fire control were superb throughout the fight – which was officially called the Battle of Gubat but known to the men as the Fight to the Nile. When the sun had set, we could hear the women and men searching for their relatives on the battlefield. Our casualties during the march of the square were about 50 killed and wounded. Some of the war correspondents marched with the square and others remained in the zereba. The necessary outposts were thrown towards Metemneh and along the riverbanks before the troops were able to rest. This was the first chance of sleep I had since the night of the 15th.

'At daybreak on the 20th, we stood to and, leaving the wounded in charge of the Heavies, the Guards and mounted infantry marched back to the zereba, carrying as much water as possible in mussicks and water bottles. In passing over the previous day's battlefield, it was noticeable that the enemy had practically removed all their dead and wounded but did not attempt to interfere with our march. On sighting the zereba we were received with such cheers as amply repaid us for the previous day's work. The zareba had not been interfered with during the night. The progress of the fight had been seen by the lookout posts up to the time of the enemy charge, but when the square had topped the slope and moved down towards the river, no more was known, except that the night was quiet and that was taken as a good omen. After breakfast the column prepared to march back to the river. A fighting square was formed inside of which our commander, Sir Herbert Stewart and

Captain Crutchley of the Scots Guards, were carried on stretchers by natives. Another square was formed of led camels, with all our wounded and supplies in the centre, and this square marched in echelon to the right rear of the fighting square. Each wounded man was given a bottle of water and made as comfortable as possible.

'Our return began at 2pm but was not interfered with by the enemy and we arrived at our previous day's bivouac at 5pm The Heavies had been busy making everything secure and had put several of the houses at Gubat, which had been vacated by the enemy, in a state of defence. These houses and the village of Gubat were occupied by the Guards Camel Regiment. The remainder of the column, including the wounded, guns and stores, bivouacked on the Nile bank.

'At daybreak on 21 January, we formed up in column of regiments (dismounted, camels left in camp) with only ammunition camels and moved towards Metemneh, covering our movements by cavalry vedettes. We were between the river and Metemneh but failed to draw the enemy. When near Metemneh two squares were formed: one composed of Guards and the other of mounted infantry. Still no sign of the enemy. On reaching the high ground the enemy flags could be seen, but it was plain to all that the enemy intended to fight from behind the walls and not try another charge. The squares, still moving with an interval between each, gradually inclined towards Metemneh. When about equal distance from the river and Metemneh, the enemy opened fire from two Krupp guns, which had been silent up to this moment, and dropped one shell in our square but, fortunately, only killing an ammunition camel. The shot was a surprise, as up to this we had no knowledge that the enemy had any guns a Metemneh. On the other hand, we should have realised that as Gordon's steamers from Khartoum were continually raiding up and down the river, Metemneh itself would be fortified on the river side against this raiding. Probably had we approached from the desert side we should have found less resistance.

'The enemy's riflemen also opened fire on us, and for a few moments, it was not at all pleasant. Fortunately for us, the enemy were very indifferent gunners and riflemen and the gunfire, after the first shots, went over our heads. The square was halted and the front, right and rear faces brought into line with the left. When formed in line we lay down and commenced volley firing by divisions and, during which, the remainder of the force withdrew behind the sandhills. When this movement was completed, the mounted infantry retired very slowly by companies and re-joined the column. During the time we were engaging the enemy, four steamers were sighted coming down the river and were met by Mr Bennet Burleigh, the war correspondent of the *Daily Telegraph* who brought back the news that they were four of Gordon's steamers with eleven guns and 500 men under Hashm-el-Mus, one of Gordon's Nubian officers. They landed five guns and with our guns shelled Metemneh until 2pm when we returned to camp. The enemy was determined to hold Metemneh and, as we were only about 900 all told, it was decided to await reinforcements before trying to carry the town. Most of the reinforcements, which had arrived from Khartoum, were most unreliable Egyptians and only their commander and about fifty Nubians could be depended on to do their work. Gordon, in a special letter to the commander of our column, requested that the ship's crews and Egyptian should not be sent back, but replaced by our own men. We moved back to our morning bivouac and at once started to put the ground in a proper sate of defence, making a temporary breastwork of camel saddles, ration boxes and anything else that could be found. The Guards completed the defences of Gubat, which was only a short distance from the river, into a fort they called it Guards Fort.

'Next day we started the permanent river entrenchment, which ran parallel to the river with an outer ditch 15ft deep and the two sides extending at right angles from this down to the river. The river side was looked after by the converted river steamers. At the desert side of the entrenchment, we had two screw guns in each corner and two Krupp guns covering the road from the Guards Fort to our entrenchment and on the river. Outside the main entrenchment and on the Metemneh side, the cavalry and horses were in a redoubt called Hussar Fort. The entrenchment was manned as follows: Heavies and two guns on the right front; mounted infantry and two guns on the left front and left face; the Sussex detachment on the right face. The hospital and commissariat were inside the entrenchment. The hospital had a few small Indian tents and what wounded could not be accommodated in the tents were placed on stretchers on the ground and protected from the sun by temporary shelters of blankets and waterproof sheets. The enemy did not attack at night, but kept up a continual beating of tom-toms, which had a very disturbing effect on the wounded.

'The clothing in use by the column was, if anything, rather warm for the extreme heat during the day in the Sudan, but perhaps better than drill uniform. The nights were dry and cold, and the light grey serge and breeches were then most appreciated. Our rifle, the Martini-Henry, was well finished and with great stopping power. The breech-falling block had to be protected by some wrapping to prevent the entry of sand which would stop the block working. The cartridge was unfortunately wrapped in rolled brass, instead of solid cases, and this, owing to the brass being crushed, was the cause of many jams during the fights. The bayonets were, in most cases, manufactured in Germany and most unreliable and brittle. It was after this war that the Small Arms Department commenced their yearly inspection of rifles and bayonets and so put a stop to any such scandal in the future. Our screw guns were easily outranged by the enemy guns. In the South African war, and certainly at the beginning of the Great War, our field guns were inevitably outranged by the enemy's guns. The enemy was in possession of the Remington rifle with a pullback bolt and cartridges in solid brass cases and not liable to damage or crushing. Fortunately, as already stated, they had very little knowledge of sighting or wind.

'It must be remembered that the column had been practically three weeks on the move, had suffered severe losses, forced by want of water to practically live on soup made from soup powders, biscuits and only very small quantities of tinned meat; added to which sleep had been denied them owing to night marching and close enemy fire for days at a time. The column was undefeated and ready to undertake any further effort that might be necessary and also, in accordance with orders issued, which were to establish ourselves on the Nile, send forward a small party to show itself at Khartoum and await the River column, which was working up the river. We had about fourteen days' provisions for all ranks and our nearest supply depot was at Gakdul Wells, about halfway between our present position and Korti. The camels had to depend on what grazing could be found.

'The ration of water on the march, not at wells, was one pint for tea before marching, not quite a pint for the march, and one pint for tea when halted at night. We had tins of Erbswurst soup powder, which was very sustaining, the water had to be found from the evening supply for tea. The tinned meat was in 6lb tins, but only sparsely used at night owing to the thirst set up. No fresh meat, vegetables or bread could be obtained during the three weeks desert marching. The camels could only be watered at Gakdul, which meant that at the most they had only three waterings in a period of three weeks and none for the last ten days. Fully 20 per cent of camels must have dropped out on the march and had to be shot or left, and those that reached the Nile were not fit to carry on; and, in fact, never recovered their condition to the end of the campaign. The camels' corn and green forage rations on the river were good, but on the desert march it meant merely giving them what could be spared. The cavalry horses, about a hundred, were in the same position, but it was possible in this case to allow a small quantity of water to each horse from the men's supplies. The officers and men on the riverbank made shelters for themselves by digging into the bank and during the day, when not on duty, could move about in the vicinity of the entrenchment.

'Metemneh was about 3,500yds from our position on the Nile and Shendy, on the opposite bank (which was held by the enemy) was about six miles downriver. General Stewart was taken on board one of the steamers. The steamers were cleared of all natives except the actual crews and taken over by Sir Charles Beresford and his naval personnel, who had the vessels thoroughly cleaned out and refuelled with wood and made ready for further service.

'On 22 January everyone was early at work in putting the defences in good order. Stock was taken of our provisions and ammunition. At 7:30pm the same evening, half of the fighting force started a march back to Gakdul Wells in order to bring back a convoy of provisions for the force. With this force was sent Captain Piggott, the Commander of D Company, mounted infantry, who was entrusted with the Desert column commander's despatches of the fighting so far. He was instructed to push on to Korti as soon as he thought it was safe to leave the column. He left the column the same night. I did not go with this column, but the Royal Scots Fusiliers did. The war correspondents, except Bennet Burleigh of the *Daily Telegraph* and Williams of the *Chronicle*, left for Korti with the convoy. As far as we know, the enemy was not aware of the convoy's departure; if they were, it made no difference to those behind, and no attempt was made to interfere with us.

'On 23 January, Sir C. Wilson, Lieutenant Stuart Wortley, two officers and thirty men of the Royal Sussex Regiment (dressed in red serge), and also two signallers, left in two steamers for Khartoum so as to show the garrison at Khartoum that our forces were near. The idea was to disembark the men at Khartoum, march through the town and then reembark and join our column. This was meant to hearten the defence.

'In the interval between 23 and 30 January the remaining two steamers were used for raiding purposes, and during this the mounted infantrymen were employed on this duty. We raided the town of Shendy which, as previously mentioned, was not far from Metemneh and near our camp. The enemy cleared out on our approach, but we obtained a lot of dhura *(sorghum)* and also a number of sheep and goats. It was arranged that we should have the sheep and the goats would be given to the Sudanese and Egyptians in our service.

'The convoy we sent to Gakdul on the night of the 22nd returned about noon on the 30th but did not by any means try to hide its arrival. It was in our interests for the enemy to think we were receiving reinforcements. The convoy brought back thirty day's provisions. We heard that the native sent in to Korti by Mr Bennet Burleigh, the war correspondent, after the Battle of Abu Klea, had reached there in good time, but as no official despatch was received until Captain Piggott arrived, all sorts of rumours were set going for about ten days by the home papers. Some thought we had been destroyed, others we had reached the Nile but were beleaguered by the Dervishes – this was pretty near the mark – and others that we were through to the Nile and pushing on to Khartoum.

'Just before daybreak on the morning of 31 January, Lieutenant Stuart Wortley and four men arrived by boat close to my dug out and brought the news Khartoum had fallen the day before the arrival of the steamers on 27 January, and that on their way back the native pilots had run the steamers aground and then deserted. Gordon was killed as he left Government House. The news cast a gloom over everyone and added to this, our loss had been about 150 killed and 200 wounded out of a fighting force of less than 1,100. This put a stop to any idea of pushing onto Khartoum until we were reinforced or the River Column arrived.

'On 1 February two Nile steamers under Lord Charles Beresford and his naval detachment, together with twenty men of the mounted infantry under Lieutenant Bower of the 3rd/60th Rifles, proceeded up the Nile so as to affect the release of Sir C. Wilson and his party. The Nile steamers were the ordinary small river steamers that could be seen plying up and down the Thames at this period. They were armed with small Krupp guns and protected from rifle fire and other fire by means of railway sleepers and sandbags. The same night we despatched our second convoy to Gakdul and apparently it got away without any trouble from the enemy. We were anxiously awaiting news from the Khartoum party. Lightly wounded accompanied this convoy.

'On 4 February the steamers under the command of Lord Charles Beresford returned, bringing with them Sir Charles Wilson and his party. The relieving steamers engaged the enemy's fort on the Gubat side of the Nile. A shot from the fort had reached the steamer's boiler when just in front of the fort, and as steam had to be cut off during repairs and the vessel anchored, the mounted infantry on the steamer prevented the Dervishes from using the gun again. The repairs lasted throughout the afternoon and night. In the meantime, Sir Charles Wilson's party, having heard the firing, marched down the opposite bank and made contact with the steamers without loss. The casualties on the relieving steamers were 1 killed and 2 wounded.

'On 11 February our convoy from Gakdul returned with Sir Redvers Buller and the 1st Battalion Royal Irish Regiment. The Royal Irish had marched from Korti to the Nile across the desert, a distance of nearly 200 miles. The Regiment were the winners of Lord Wolseley's Cup for the best work under the prevailing conditions in getting the whaler boats over the cataracts and up to Korti. The men looked very fit but inclined to be fine drawn.

'It was rumoured that we were to storm Metemnuh on the Sunday. We heard that the River column had been successful at the Battle of Kirbekan, but that General Earle and Colonel Eyre had been killed. On 13 February the remainder of our sick and wounded left for Korti, being carried on stretchers by Egyptians and Nubians and those who could sit up, on camels. Sir Herbert Stewart was amongst the wounded and the convoy was escorted by the Heavy Camel Regiment. The enemy got in touch with this convoy but was beaten off and the convoy reached its objective in good time.

'That evening all surplus supplies were burnt or destroyed. Gordon's steamers were disabled, and the guns thrown overboard. These same steamers were repaired and used by the Mahdi during the Sudan campaign of 1898. All the troops encamped outside the fort on the night of the 13th, except D Company of the mounted infantry inside the fort. The men had a bottle of wine between each six men from the surplus stores. At daybreak on 14 February, the remainder of the column, consisting of the Guards Camel Regiment, Mounted Infantry Camel Regiment, Royal Irish Regiment and a detachment of Hussars left Gubat for Abu Klea, on the return journey all, except the Hussars, being dismounted. What camels we had were not fit to carry anything beyond about one cwt. of supplies or kits and could not be used for men who were worn out, except to have them hang on to the girth straps and stirrups. The march to Abu Klea was slow and deliberate and about 12 noon we halted in a good commanding position in the desert, with the hope that the enemy might be encouraged to attack, but nothing happened, although the enemy's scouts kept in touch with us as far as Abu Klea.

'At daybreak on 15 February, we continued to march to Abu Klea and arrived during the afternoon. The Light Camel Regiment were there. They had been on escort duty between Korti and the wells since the beginning of January. Captain Kitchener was in charge of intelligence at Abu Klea, as Sir Charles Wilson had left for Korti. Kitchener was hardly likely to be again dispossessed of the appointment. The Heavy Camel Regiment had gone on to Korti with the wounded and did not re-join us again. We were all in a deplorable state for clothing and boots and particularly so in the mounted infantry, as we had never been near any Ordnance Depot after leaving Wadi Halfa in October. The men were, in many cases, without boots of any sort and marched with their puttees wrapped round their feet and legs. The authorities at Korti were asked to send boots by convoy. Although the enemy scouts had been seen at Abu Klea since its occupation a month previous, no attempt had been made to attack the position. Several of the enemy wounded from the battle on 17 January were now at the wells and being attended to by our medical officers.

'On 16 February the squadron of the 19th Hussars and Guards Camel Regiment left for Gakdul and Korti. The Abu Klea force sent out Cossack posts to the surrounding hills, so as to cover the wells. The force consisted of Sir R. Buller in command, the Light Camel Regiment under Colonel Sir Stanley Clark, the Mounted Infantry Camel Regiment under Major Barrow, 1st Battalion Cameronians (Scottish Rifles) and the 1st Battalion Royal Irish Regiment. Lord Charles Beresford was in command of the Naval Gun Detachment. The Light Camel and Mounted Infantry Camel Regiments were in one irregular shaped camel saddle breastwork. The Royal Irish Regiment was in a breastwork on somewhat higher ground than ours, and the Naval

Detachment in a small fort between us and the road to Metemneh. This fort was also, I think, used afterwards for wounded, and the gun positions changed. We used the natural nullahs and camel saddles for shelter. Our position on the north side was commanded by the Royal Irish position and we, on the other hand, were on our north side commanding the Royal Irish south side and on our west side, commanding the commissariat and naval fort. When these dispositions were made, I presume no attack was expected from the enemy. The wells were only giving a very limited supply of water, and none could be spared for the camels, but as the wadi had plenty of course grass growing in it we were able to take our camels out and graze for some hours. At about 3pm the same day, news was brought in that a considerable number of Dervishes were coming from the direction of Metemneh. At about 5pm the enemy attacked the surrounding hills and drove in our posts to the wells. The enemy now held positions which overlooked ours and commenced a heavy rifle fire. We could not reply as the enemy were well sheltered by rocks and hills.

'Early that evening our Regimental Sergeant-Major Harwood was killed and Captain Walsh of the Somerset Light Infantry was shot through the lung: he fortunately pulled through in the end. This long-range firing was most trying. Our screw guns must have left on the 15 February with the Guards Camel Regiment as I did not see them again, nor were they in action during this attack. After dark we formed round our camels and prepared for an expected attack, but nothing happened. The Royal Irish Regiment did have a false alarm about 10pm but fortunately their fire was soon controlled. My orders from the officer Commanding D Company were to fight towards the fort with the Royal Scots Fusiliers in case of attack. If an attack had been driven home that night, we simply stood no chance and would have shot each other down in the melee and, in any case, how was the garrison of the fort to discriminate? During the night, the enemy kept up a continuous fire and we had many casualties. We looked forward to daybreak with relief to our nerves.

'At daybreak on 17 February, the enemy rifle fire increased in intensity and, when going on duty at the wells, two of our men, Lance Corporal Fraser and Private Monaghan, were hit. I went to see them later in the temporary hospital and found that Fraser had been shot in the mouth and Monaghan in the hand. Although Fraser could hardly speak owing to his wound, he had a joke about Monaghan's wound. He said Peter would not have been wounded if he kept his hand steady as ordered. At this time in the 1st Battalion, the Sergeant-Major would have the disengaged arm kept perfectly steady at the side, hence Fraser's remark. Monaghan was very sorry for himself and did not quite see the point of the joke. At about 2pm Major Wardrop, Lieutenant Tudway of the mounted infantry, and three men of the Hussars made a reconnaissance towards the hills and when close to the enemy position, the enemy cleared off at once evidently fearing an attack. We occupied the positions without delay and so ended our nerve-racking experience of continual rifle fire without return. Strong positions were taken up on the hills so as to cover the wells and our bivouac, and we then carried on the usual daily duties.

'Camels were sent into the wadi daily to graze under a strong escort of mounted infantry. One morning I went out in charge of this party and when well in the wadi, two Arabs jumped up just in front of my camel and then dropped on their knees, holding my stirrups. The interpreter said they were deserters from the enemy. They were kept under guard until evening when we returned to camp and handed them over to Captain Kitchener. A couple of evenings after, when returning with the grazing party, we met the two Arabs coming in our direction with three Bashi-Bazooks as guard. By the time we arrived in camp, the Bashi-Bazooks had re-joined us, and I was told that the Arabs had been shot as spies. The secret I heard that evening. It appears that when the Arabs were handed over to the Intelligence on our return to camp, they were confined separately. During the same evening, a deserter came in openly and when questioned said he had deserted and that the evening before two men had been sent out to find out our strength and position. He was confronted with our prisoners and identified them. They were sentenced to be shot. Unfortunately for the spies, our grazing party that morning had gone off the beaten track and they could not clear out of the way and evidently thought it best to show themselves as being less likely to cause suspicion.

'My diary finishes here, so hours and days have to be missed.

'We remained at Abu Klea until about the last week in February. The strength of the enemy was gradually increased by reinforcements from Khartoum. During our last afternoon at Abu Klea, we, the transport and supplies, received orders to move at once. We got off early in the afternoon and marched about six miles back on the road to the Gakdul Wells, in good open country, and where we found a small post with a supply of water in iron tanks (ten gallons each) to meet requirements of the retiring column. Similar posts were placed on the main route to Gakdul and were of great assistance, as nothing could be saved from the supply at Abu Klea.

'The main body left Abu Klea after dark, leaving all fires burning and carrying the enemy wounded with it as far as the first halt. The enemy wounded had been treated well, but usually spent the night shouting out information in case any of their own people were near and would certainly have given away the fact of our departure. Next morning when we marched, the enemy wounded were left behind to be taken over by the Dervishes, who were in touch with us within a few hours of our departure and followed up until our arrival at Gakdul. On arrival at Gakdul we first heard of Sir Stewart's death. He was buried under the shelter of the hills and close to the wells. A fine, gallant soldier and born leader of men. Lieutenant Rycroft of the Highland Light Infantry, joined on this march and took over command of the Royal Scots Fusiliers Division. Sir Evelyn Wood was in command at Gakdul, Sir Redvers Buller having left for headquarters of the Expedition at Korti.

'The Gakdul Wells had many pepperbox posts on the surrounding hills overlooking all approaches to the wells. We received here about 200 pairs of all sorts and conditions of boots for our men; it was a consignment that had evidently been discarded by others on receiving new issues. We made the best of them. I usually take a size 6 and had to manage with a twisted pair of size eights. I know when I arrived at Korti later my feet were in perfect condition and any sign of a corn that previously had been noticeable, was completely removed by friction. The road from Abu Klea to Gakdul was, on the whole, level going, but at times we struck a patch of deep sand, about 12in deep, for two or three miles. Our longest march was twenty-nine miles on a day when the enemy seemed to be pressing us hard. The stench from the dead camels on this route was almost unbearable. We must have lost hundreds during the three months we had been in the desert.

'Gakdul itself we found beautifully cleared of rocks etc. water supply improved and generally with a prosperous appearance. The day after our arrival a big convoy of donkeys and camels arrived, and we were able to give our poor camels a good feed. We did not ride them at any time on the return journey but tied them in threes from head to tail. After a couple of days at Gakdul the mounted infantry received orders to start their march to Korti the following day. We were delighted. Next morning the outposts reported a large number of the enemy about two miles from the wells. Our march was cancelled, but the remainder of the force moved off, leaving Sir Evelyn Wood and the mounted infantry to cover the retirement to Korti. Our retirement was intentionally slow and deliberate and, although we were watched all the way, no attempt at an attack was made. When we reached that part of the desert where we had captured our first prisoner – the bandit. He was released, given some dollars and quite good supply of rations. He had been attached to the mounted infantry for safety and rations and I had often talked to him through our interpreter. When he left, he was most profuse in his thanks and wished me all sorts of good luck. He retained his donkey and said he knew where his wife would be and that he could easily keep out of the way of any Dervishes he did not want to meet.

'The main body had arrived at Korti some three or four days before we did, which was in March. As we retired, we withdrew with helio signallers at the fifty-mile wells and thirty-mile hill. We arrived at Korti after being through a most trying three months in the desert, subsisting on minimum of rations, with very little water. The morale of our men was splendid, and I do not remember a case of ordinary sickness throughout this period.

'The Commander in Chief of the Nile Expeditionary force, Viscount Wolseley, inspected the mounted infantry the day after our return. He thanked the Regiment for its splendid services and told us we should have a post of honour on the march and entry to Khartoum in the autumn. Unfortunately for us, this march did not happen for over thirteen years afterwards. At Korti we found that the troops of the expedition were going into summer quarters on the Nile, with the advance base at Merawi. We, of the mounted infantry were to be split up into two half Battalions – one at Tani and the headquarters and the other half at Kurot.

'The Mahdi's force made no attempt to interfere with us during this period, but this might be due to the fact that another force was operating from Suakim in the direction of Berber and was attacked with great determination by the Dervishes. In May, rumours of withdrawal were very prevalent, and these were confirmed. Our orders were to proceed by native boats (nuggars) to Wadi Halfa and await further orders. When we arrived in Cairo orders were received for all detachments to re-join their Battalions forthwith, and I saw the two Companies for England.'

Chapter 4

Establishment of the mounted infantry and operations in Mashonaland and Northern Nigeria

The splendid work done by the MI in the Egyptian campaign of 1882 had caught the imagination of the general public, but there was much professional opposition to the proposed scheme for a permanent mounted infantry arm. Lord Wolseley, who commanded the Egyptian campaigns of 1882 and 1884 and employed MI units raised on an ad hoc basis, was a strong advocate of this arm and held the influential position of adjutant general to the forces from 1882 to 1890. In 1881 the issue had been reviewed by the Intelligence Division, who considered the MI's military role, how it would interface with the cavalry and whether it should be established on a permanent or impermanent basis. The report provided three options for the future of the MI:

1. A permanent corps.
2. Maintaining the current system of raising of units from locally based infantry regiments.
3. Producing a cadre of peacetime trained men abstracted from infantry regiments and available for deployment when required.

Option one was not taken up, though the question of permanence or impermanence remained current for the rest of the MI's existence and centred around the arguments of available finances and manpower. The Secretary of State for War was asked in November 1882 whether a permanent nucleus of mounted infantry would be established to avoid the current method of improvising on an as required basis. His response was, '... to keep on record the organization requisite for such a force and to maintain the necessary equipment in store, it is not deemed advisable, with our limited establishments, to create the nucleus of a force of mounted infantry.' Consequently, option three was favoured over option two.

In June 1886 Major Edward Hutton, a leading authority on the MI, presented a lecture entitled 'Mounted infantry'. Lord Wolsely chaired the meeting attended by senior officers from the infantry, cavalry, artillery and volunteers, wherein he first raised the principle of raising and training a force of mounted infantry. He argued that the object of the mounted infantry was to provide an improvised substitute for cavalry in small wars and the necessity for some previous training and a general system of organisation for mounted infantry. Wolseley concluded the discussion that followed Hutton's lecture by saying, 'Mounted infantry as a force in the field is of incalculable value. My idea of cavalry and of mounted infantry is, that they are almost two distinct arms of the service.' Hutton recognised the influence of Lord Wolseley's 'intuitive military genius [that] we owe the establishment of mounted infantry in the British Army'.

The question of whether mounted infantry work should be handled by cavalry or infantry was argued from both sides. Cavalry supporters maintained that cavalry could be trained to take the duties of the mounted infantry and in doing so would be a more effective force. Major Hutton accepted this view but argued that the current size of the cavalry was not sufficient to provide an adequate force for all expeditions and areas where British troops were stationed. In addition, Wolsely expressed his opinion that the cavalryman was intended to fight on horseback and to make him fight on foot in an infantry role was a waste of power. The mounted infantry concept represented a cheaper option for carrying out certain mounted duties such as vedette and outpost work, orderly and escort duties, leaving the highly trained cavalry troops and horses to fulfil their designated roles in reconnaissance, raids and engaging the enemy's cavalry in its traditional *arme blanche* role.

A cavalry soldier and his mount required extensive training, over a period of around eighteen months, to come up to standard and be able to fight effectively from the saddle. The training requirements of the mounted infantry were focused on horsemanship and horse management, which could be provided in a short space of time to bring them up to the standards required. The MI soldier was already trained as an infantryman, eminently suitable for the dismounted fighting role assigned to them. His equipment was suited for mobility and dismounted action; the cobs or ponies he was mounted on were smaller than cavalry horses and this enabled him to mount and dismount quickly as opposed to the heavily accoutred cavalry man. He was equipped with the latest infantry rifle that fired with a higher muzzle velocity than the cavalry's carbine, resulting in a more powerful weapon with a longer range.

The recent campaigns on the African continent had relied on locally raised and improvised mounted infantry units drawing infantry volunteers from the line regiments taking part. Supporters of the mounted infantry were only too well aware that there was a pressing need for a more structured method of raising and training mounted infantry. The question then arose of whether this should be a permanent force benefiting from a defined clarity of its role and training, regimental identity and identifiable promotion prospects for the officers and men. The counter arguments were that a permanent corps would, over time, shift to becoming an undertrained, indifferent cavalry, losing the infantry capabilities without gaining the shock tactics skills of fighting in the saddle. This last point was refuted strongly by Lieutenant General Sir Edward Hamley, who maintained that this would be held in check firstly by their own officers, secondly by the Inspecting Generals and lastly by the authorities responsible for establishing the corps.

The alternative (and adopted for peacetime) was a force based on impermanence whereby officers and men were abstracted from infantry regiments and trained for an MI role, returning to their units when the training course had been completed. This cadre was then available for mobilisation and deployment as required and was seen as way of keeping costs down and reducing demands on the budget to finance extra cavalry regiments. Although this found favour among senior officers, abstraction of men from infantry regiments was unpopular with the infantry commanders who lost experienced soldiers, albeit on a temporary basis, for training and possibly longer periods if deployed on mobilisation. They further considered this would have a detrimental effect on regimental discipline and *espirit de corps* by removing these experienced officers and soldiers. Lord Wolseley disagreed; he considered the regimental spirit was strong and resilient enough to ensure that the removal of a couple of good officers and a party of men would not destroy the battalion. 'We have plenty of good regimental officers and there is no battalion in the Army that could not spare two or more officers for mounted infantry work and spare them well.'

Lacking a formal doctrine, the MI relied on certain general principles together with drill and training manuals for guidance; however; this did not provide standardisation or uniformity in its application. While the MI's exact role when deployed in the field may not have been defined, that of the mounted infantry man was. He was, first and foremost, a mobile infantryman who would fight dismounted; the cob was for locomotion purposes only. The question remained whether they should be attached to the cavalry, providing them with an infantry capability as the Royal Horse Artillery provided artillery support; or to the infantry as a mobile reserve; or as an independent mobile force. The manuals generally disavowed the idea of a cavalry function, emphasising firepower and mobility strength that should be applied in scouting and providing protection for infantry, cavalry and artillery. The establishment of a training school would go some way to disseminate a standardised MI.

Hutton's proposal for training was to establish at Aldershot a school and cadre corps of mounted infantry. Infantry battalions were to send small detachments for a three-month training course covering the duties of mounted infantry, principally focusing on the care of horses, riding and stable management; movement of mounted troops in formation and in the field including the advance and withdrawal when fighting; patrols, outposts, observation and extensive reconnaissance. The school itself, consisting of a limited number of officers and NCO instructors, would be the nucleus upon which a large corps could be embodied. This was longer than the three days' equitation training stipulated in the 1884 Regulations for mounted infantry and in 1888 the first mounted infantry School of Instruction was established by Major Hutton at Aldershot.

Hutton's first adjutant was Captain Edward Alderson[8], who, in 1896, was appointed to command four companies of MI for service in Mashonaland.

---

8 Later **Lieutenant General Sir Edwin Alfred Hervey Alderson, KCB;** born 8 April 1859, died 14 December 1927. Served in the First Boer War in 1881, 1882 Anglo Egyptian War and two years later in the Mounted Camel Regiments expedition to relieve Gordon at Khartoum. He was then appointed to the MI Depot at Aldershot before going out to Mashonaland in 1896. He returned to South Africa in 1900 commanding Mounted Troops.

These companies consisted of four sections, each section abstracted from an infantry battalion. Alderson explained how, in his opinion, the 'scratch patch' he commanded worked well because of its organisation. Each section comprised approximately thirty NCOs and men and one officer. The officer commanding an MI company was selected by the Commandant of the MI in Aldershot for approval by the relevant authority. The principle on which the MI section, company or battalion was worked, was one of decentralisation. The men in the section organised themselves into permanent subsections of four who messed and conducted duties together and selected their own leader who was responsible for the subsection. Their horses would always be piquetted together and, when in action, one soldier was responsible for looking after the four horses whilst the other three operated dismounted, in the infantry role. In 1902, Alderson was Inspecting General of MI in Natal and wrote to Colonel A. Godley (who had been appointed in 1901 DAAG on the HQ staff at Aldershot, which carried with it command of the School of Instruction for mounted infantry) outlining his views on the organisation of the MI based on his recent campaign experiences. He stated that he had been converted to be in favour of the MI company being formed from one infantry battalion because of the difficulty experienced of keeping a company made up of sections drawn from a number of infantry battalions up to strength. While originally favouring abstraction from one infantry battalion to form a section because, 'As a rule [it] gives you better officers and men.' He acknowledged that having more than four infantry battalions was proving fatal to cohesion and efficiency as it, 'Does not stand up to the test of a long campaign' and leads to 'split sections'.

Mashonaland 1896

Mashonaland (modern-day Zimbabwe) was occupied without resistance by European settlers trekking up from South Africa looking for the second Rand in 1890. Settlers were allowed to choose farms where they liked, though land was not to be taken where it was in use by the natives as their village and cultivated fields. Expectations were quickly dissipated and the British South Africa Company shares began to fall as many pioneers returned to South Africa. The Shona felt little pressure on the land from settlers and some deference was afforded to the Shona, who remained in control over the greater part of their tribal land. However, they were subject to the annual hut tax of ten shillings and the general brutality of the age.

Cecil Rhodes now looked to Matabeleland, considered by him to have similar reserves of mineral resources. In 1893 the Victoria Incident involving the murder of white settlers gave the British South Africa Company the excuse to invade Matabeleland and defeat the Ndebele army in the First Matabele War. This result was to open up the land to white settlers to acquire farms without, however, first studying the consequences for the native population. To legalise the move a Land Commission was formed to look into the issue and reported in October 1894, taking the least line of resistance to the British South Africa Company and white settlers. The high *veld* was occupied by the Europeans and two areas of 3,500 and 3,000 square miles assigned as reserves for the natives.

Following Jameson's failed attempt to invade the Transvaal in 1895 the Ndebele took the opportunity to rise in armed conflict in March 1896, beginning the Second Matabele War. The Shona followed suit in 1896.

Initially the local white population banded together to escort outlying settlers into safe *laager*s and undertake punitive measures commensurate with their available resources. The Matabele Relief force commanded by Colonel Plumer[9] was sent out to quell the rebellion. In July a force of mounted infantry, originally sent from Aldershot under the command of Colonel Alderson to serve in Matabeleland, arrived in Mashonaland via Beira.

Captain Alexander John Godley, then adjutant of the Mounted Infantry School Aldershot, was appointed Alderson's adjutant on this expedition. Alderson recorded his experiences in 1898 in a book entitled *With the Mounted Infantry and the Mashonaland Field force 1896*, while Godley was asked to contribute a chapter recording his experiences. He suggested the inclusion of this chapter as an example of the work of the MI in that expedition for the short history being prepared by Tomlin and Craufurd for the MI Club. His account is given below:

The Sinoia and the Ayrshire Mine Patrol, by Brevet Major A.J. Godley.

---

9 Later **Field Marshal Herbert Charles Onslow Plumer, 1st Viscount Plumer, GCB, GCMG, GCVO, GBE, KStJ**; born 13 March 1857, died 16 July 1932.

'On 24 October, having seen Colonel Alderson off to Salisbury and Jenner to Mazoe, I started in the afternoon with a column for the Lomagundi and Sinoia district. The force was composed as follows:

	All Ranks
Staff	8
Scouts	12
Irish Company Mounted Infantry with Maxim	109
Medical staff corps	6
Mounted Troops Rhodesia Horse	30
Artillery Troop Rhodesia Horse (one seven pounder and two Maxims)	30
Salisbury Rifles	35
Native contingent	76
Total	**306**
With 21 mule waggons	

'Colonel Alderson had kindly lent me Wilson, Fox and Harding from his staff to act as gallopers; Lieutenant (now Brevet-Major) Pilson of my regiment, was in command of the mounted infantry company, and acted as staff officer. Captains Daly (late 15th Hussars), and Moberly (late RA), were in command of the cavalry and artillery of the Rhodesia Horse and, with Captain Brabant to lead the native contingent and Surgeon-Captain Hale DSO to look after the medical arrangements, I considered myself very lucky in my command. The only difficulties were mealies and boots. We had a long way to go in a certain time, with no mealies for the mules and uncommonly little for the horses. In the event of much rain, with consequent heavy roads and swollen rivers, it would be touch- and-go whether the animals would last long enough to enable us to visit as much of the country as the Colonel wished. The question of boots was a serious one too. Many of the Salisbury Rifles, and the dismounted party of the mounted infantry, whose horses had died, were almost barefooted, and one could hardly have wondered if they had grumbled at being now ordered to turn their heads away from Salisbury and start on a practically fresh expedition. However, so far from grumbling, everyone had made shift to get himself some sort of foot-covering. The Colonel, and one or two of the other officers who had them, had supplied lawn-tennis shoes, the sick going back to Salisbury had contributed anything they had, and the Natal Troop, who were to escort the Colonel, had taken their boots off, and exchanged them for the remnants from their foot-slogging friends. Those who had not been able to cover their feet had sewn strips of horse or ox-hide across the holes, and so we set off. We hardly started when down (without any warning, as is the way of them in South Africa) came a terrific rain and hailstorm, and, as I watched the mules straining up the hill by Eyre's Farm, and the foot sloggers toiling and slipping in the mud, I must confess to having felt rather downhearted at the prospect of a rainy trip. However, the men didn't seem to mind, and we contented ourselves with a short trek for the first day.

'We had to do a long march next day, so after a very windy and uncomfortable night (fortunately the rain had stopped) we inspanned at 3.00 am and, by nightfall, had reached and *laagered* up at a valley near the Bluecoat Mine. We had travelled all day through thick bush without seeing a sign of a *kraal* or a local. Our marching troubles had begun though. The road was very rough and sandy, we had just enough rain to make it heavy, the crossing of the Maninie River (where we outspanned in the middle of the day) proved fatal to the dissel-boom of one of the waggons, and there were four mules less at evening roll-call than had started in the morning. Still, we had done fourteen miles, and the weather was looking brighter.

'On the 26th we came upon the first traces of the murders in the Lomagundi district. Our midday outspan was near the Sange River, and there we found two skulls, some bones, and some partly burnt letters from Jameson, the Mining Commissioner of the district, who was supposed to have been murdered. Hale pronounced one of the skulls to be that of a white man, so we took it with us and buried it next day at the store.

'In the afternoon Pilson, with some mounted infantry and scouts, went on to Deary's store at the Doondu River to take over and make an inventory of the contents, while Harding and Southey with another party went to the Mining Commissioner's camp and collected all his papers and effects. I followed later in the column and, on arrival, collected all papers of importance from the parties which had gone on in advance, and commandeered and divided equally among the troops what stores had been found. The rebels had left practically nothing though but a few bottles of pickles, sauce etc. Both camp and stores were completely wrecked and looted, and there were abundant traces that they had been living in the huts. In one of them they had killed a buck, apparently only the day before we arrived, and there was fresh spoor all around the place leading into the bush. Two diaries were found in the Mining Commissioner's office, one written up to 21 June and the other up to 22 June. We either burnt or brought away papers or letters, according as they appeared important or not. We had now done only twenty-six miles, the Rhodesia Horse had finished their mealies, and I wondered how many horses would get back to Salisbury. There were signal fires at night on the Sinoia Hills, and they, coupled with the desertion of this place, made us afraid that the locals intended clearing before us, and that we should not have a fight.

'However, the caves of Sinoia had to be visited, as it had been reported that the rebels were living in them, so I decided to leave all the weakest horses and mules with most of the waggons here and make a flying visit to the caves. Our road to the Ayrshire Mine, which was another object of the patrol, branched off from here, so it seemed only an unnecessary waste of energy to drag all the waggons, etc, to Sinoia and back. We started with a few waggons, the best of the mules, and four days' rations, leaving the Salisbury Rifles, some casuals of the various detachments, and the mounted infantry Maxim to guard the *laager* and look after the weak horses. Lieutenant Elphinstone-Dalrymple of the Salisbury Rifles remained in charge.

'On arrival at the Hunyani River we outspanned and decided not to go on to the caves at once, but to rest for the remainder of the day so as to have the horses fresh for the job, and a whole day available if necessary. One of the first sights that greeted our eyes on arrival was a huge crocodile, which effectually did us out of the bathe to which we had been looking forward. The scouts had reported seeing plenty of spoor of locals on the Sinoia road, so hoped that they might be waiting for us.

'On the 28th we started early with the scouts, as many of the mounted infantry as we could find horses for, the seven-pounder, and one of the Maxims of the Artillery Troop Rhodesia Horse, and the Native Contingent, leaving twenty-five dismounted mounted infantry men and one of the Artillery Troop Maxims with a gun detachment for it, under command of Lieutenant Stockley MMP, to guard the *laager*. The drift across the Hunyani was a bad one and we were delayed by the darkness and by some of the horses being swept off their legs. Fortunately, we had no waggons to get over and, long before daybreak, were well on our way. The distance was much further than I had been led to expect, though, and when at last the guide reported that we were within sight of the caves it had long been daylight. However, it did not matter as, on two sections of mounted infantry and the Native contingent going to explore and occupy the caves, they found them full, not of men, but of baboons.

'From the spoor seen and intelligence received before starting, I knew there must be rebels in the neighbourhood, so I had the guns outspanned, off-saddled what horses were not required, and started off with the mounted infantry and Mounted Troop Rhodesia Horse in four different directions to scour the surrounding country. They burnt a lot of *kraals* which showed traces of recent occupation, and in one them found an old woman, whom they brought back. She said that all Sonia's people and all the natives of the district had "cleared" into the fly-belt towards the Zambesi on hearing of our approach, and that she had been left because she was too old to move quickly. She knew all about our fight at Mashangombi's and confirmed that report that his brother had been killed in action there, and that Mashangombi himself with most of his people and cattle escaped to Thaba Insimbi and Iron Mine Hill. She also told us that it was people from Zimba's and Sinoia's *kraals* who had murdered the whites at Deary's and the Ayrshire Mine.

'On the return of the parties, they off-saddled their horses for a bit, ate the half-day's rations which they had carried in their saddles, and we all took the opportunity of exploring the far-famed caves of Sinoia. People who have seen them both say that they are far prettier than the blue grotto at Capri, and certainly the main cave was quite beautiful. The hole goes down perpendicularly from a flat level surface and is a tunnel, narrow at the opening and widening out as you get into the cave. At the end of the tunnel is a pool of clear blue water, with a tree growing in front of it. As you go down you see its reflection in the water, which is very deep and most vivid blue. The effect is beautiful. On arrival at the pool, you see that the hole of the cave is circular, about 100ft in diameter and about 150ft high, with the same diameter right up to the sunlight. The clear blue sky appears through the green branches of the tree spreading overhead. The whole cave is composed of marble and the effect of the blue sky seen through the green branches above, the clear blue water below, and the surrounding walls of marble, is more lovely than I can attempt to describe. On the far side of the pool are black and gloomy caverns, which appear to lead into the bowels of the earth.

'We returned in two parties, the guns with a section of mounted infantry as escort going by the road, and the rest of us taking a short cut through the bush. The former party saw a few armed natives running away into the bush about 500yds from them but the horses of the escort were so done up that they could not catch them. We had to leave a good many horses on the road before we got back to *laager*, but they were recovered in the cool of the evening. All the horses which had been out round Sinoisa's were very much done up, chiefly, I think, from want of water and the great heat. We had not been able to find any water for them between Hunyani and Sinoia's, and at the latter place they could not get at it.

'Next day we trekked back to the Doondu, leaving behind the bones of three of our best mules, one of which had strayed and been taken by a lion the previous night. On arrival at the store the *laager* there was broken up, and the force, again re-united, trekked a short way along the road to the Ayrshire Mine. Two horses and two mules died, however, and at the rate they were going it seemed as if we should not have many to show on our return to Salisbury.

'All this time we had excellent shooting, and the scouts and flank guards had pretty well kept the column supplied with fresh meat. One of the scouts had shot a beautiful water-buck, and Southey and King-Harman, two mounted infantry subalterns, had been very successful among the smaller buck, guinea-fowl and partridges. I forget whether it was on this patrol, or before we left the Colonel, that our bill-of-fare for dinner at the headquarter mess one night was: Venison soup, barbel (a very good fish that we used to blow out of the rivers with dynamite, no time for catching them by more legitimate methods), roast eland (the best eating buck in South Africa), roast guinea-fowl, omelette of ostrich egg (someone had picked one up) and tinned peaches. Could any gourmand want more, especially when washed down with a little of our precious stock of *dop* (Cape Brandy)? I am bound to say, though, that bully beef and biscuit was more our common fare.

'30 October found us at Ayrshire Mine after a hard march. We had a rainy, windy night, and, starting at 3.00 am in rain and cloud, had made little progress at first. This was chiefly owing to the thick bush, the difficulties of keeping to the so-called track, and the necessity for incessant halts to clear the bush and cut down trees which obstructed the waggons. We had no moon now either, but the transport riders and conductors were unanimous in their opinion that, if we were to do the job with no food for the animals, we must do all the marching possible in the cool of the evening or early morning and give them the whole day to graze. The day at Sinoia's when we had been out for so long in the heat of the sun had knocked the horses up more than all the rest of the marching we had done put together. Two more of them and a mule had succumbed to this march, but when we outspanned, we had the satisfaction of finding that we had done 12 miles and had unexpectedly arrived at the Native Commissioner's camp.

'In the afternoon I went with a small escort to the Ayrshire Mine, while Hale and Fox explored the Native Commissioner's camp. We found the store and all the huts at the Ayrshire looted, nothing of any value left. The mining camp is quite a large colony of well-built huts and must have been most comfortable. It was very sad to see everything so completely and wantonly wrecked, the machinery being apparently the only thing which the local natives had left untouched. One of the scouts, who had heard stories of untold gold and treasure having been thrown down the main shaft by the fugitives after the outbreak, lowered himself down with a rope, but found there was too much water to allow investigations, and our only spoil eventually consisted of some specimens of quartz and some tobacco, cigarettes and cocoa, which my escort divided amongst themselves. The tobacco was a godsend, as most of the column had been reduced to smoking tea or not smoking at all.

'On our return, we found that the party to the Native Commissioner's camp had found and buried skulls and bones of two Europeans which were supposed to be those of Maynhardt, the Native Commissioner, and Carr or Suter. They had also found and brought away a quantity of papers. The Native Contingent had, in the meantime, been spending the day in scouring the surrounding country, and had burnt four deserted *kraals*. They had not met any rebels but had come across a lot of lately planted mealies and a great deal of fresh spoor, and, in following this up, had seen a party of natives on the top of a high *kopje* about 800yds from camp. These were seen again from the *laager*, just before sunset, so Moberly fired a few rounds of common shell at the place, which apparently had the effect of dispersing them.

'Owing to some of the men and horses having had such a long day, and to the indications of rebels about, we did not start until daybreak the following morning. Our way now lay towards Salisbury again, and along the main road from the mine. The road was very sandy and heavy and the bush very thick, and we could find no water till we struck the Mannie River again, having had to leave five horses, twenty-four mules and one waggon, on the road. Most of the mules were retrieved in the evening. The only way we could get the horses back to Salisbury now was to take the men off their backs. So, I had nearly everyone dismounted, most of the saddles put on the waggons, and the horses driven like a flock of sheep. It was a most melancholy sight to see the poor brutes tripping over every little stump or rise in the ground, then falling and simply unable to get up from starvation or exhaustion. Among them, one of mine had to go, which I had hitherto ridden more than the others. The flanking parties, on coming into the outspan, reported having burnt some deserted *kraals*, and found a complete skeleton of a Zambesi boy. They had seen a good deal of spoor but no natives.

'On Sunday, 1 November, the scouts found some very good grazing a short way off the track, so we had a day's rest and sent all the horses and mules there. We were glad of plenty of time for a good wash and what change of clothes we could muster. A great part of the day was spent in trying to get the horses and mule, which had fallen down, on to their legs again but, notwithstanding our exertions, when we moved the *laager* across the drifts in the evening, we had to leave three horses and four mules behind. The Native Contingent had gone out to try and find the bodies of some white men who were supposed to have been murdered near here on their way to Salisbury, and found two, which, as far as we could make from the papers on them, were those of Drysdale and Briggs. They were buried where we found them.

'Next day found us back at the same place where we outspanned on the first night after leaving Eyre's Farm, but minus five more horses and three mules, and on the 3rd, we crossed the Umvukwe Mountains, and met a convoy from Salisbury with rations and mails for us. They were Natal Troop and West Riding men, and from them we heard the sad news that poor Evans had been killed in an attack on Gatzi's *kraal* near Marandellas, the same place where Barnes ASC, had been killed before we got to Salisbury. We were very glad of the rations and boots which they brought, as may be imagined.

'Our journey back, as far as Mount Hampden, was uneventful, a good many horses and mules succumbed, some men went back to one of the outspans and found some natives there, and exchanged shots with them, without harm to either side, and one day the Native Contingent found the skeleton of a murdered Shangaan.

'On arrival at Mount Hampden, we were met by a most welcome waggon-load of mealies and some *dop*, which the Colonel had thoughtfully sent out. Just as we were *laager*ing up, a mule driver, who had gone about a mile off the road to look for a stray mule, staggered in all covered with blood. It appears that he had been attacked by some Mashonas, who jumped out of the long grass and assegai him in the thigh, inflicting a nasty wound. They ran away in the direction of Mashangombis. The Native Contingent went after them and found their spoor and some blankets, mealies etc, which they had evidently dropped but, after following up the spoor for some way, they saw it would be hopeless to come up with them and had to give up the chase.

'The scouts also came upon some fresh spoor and followed it up till they came to a brand-new *kraal* in the Mazoe direction, which showed traces of having recently been occupied. This they burnt and came back without having seen a native.

'On the morning of 8 November 8 we met the Colonel who had ridden out from Salisbury to see us, and by 10.00 am the column had dispersed to their different camps and *laager*s about the town, and I had to return, reluctantly, to my office stool. We had been on our own resources for sixteen days and during that time covered 140 miles. We gone considerably farther north than anybody else in the course of the campaign and thoroughly explored the Lomagundi district. We were, of course, very much disgusted at not having a fight but, from all one saw and heard, there was no doubt that most of the rebels in that country had cleared before us into the "fly". There were evidently armed parties of them hiding in the bush but one's chance of coming across them was practically nil.

'It had been a most interesting trip otherwise, the country being so different to most of the rest of Mashonaland. Instead of *kopje*s and comparatively open country, we had been travelling, except at the Ayrshire where there were *kopje*s, though flat country with dense bush, and the scenery at the Hunyani and Mannie Rivers, and the caves of Sinoia, was worth a trip from Europe alone to see. The only unpleasant feature had been watching the horses and mules dying daily. We had lost twenty-seven horses and twenty mules altogether. The men had a lot of hard marching but, not withstanding their bootless condition, had tramped most cheerfully and nobody could have had a nicer command, or given it up with more regret than I did.'

Northern Nigeria, 1897; 1900-1903

West Africa, and the ingathering of the Fulani Empire of Northern Nigeria, in the opening years of the twentieth century was, in spite of colonial administration, little known at the time. Among the many spheres of influence resulting from the partition of Africa at the Conference of Berlin in 1885, that of Nigeria fell to Great Britain.

By a fortuitous piece of good fortune, when France and Great Britain arrived on the scene, the two great black empires of the middle and western Soudan were in the process of dissolution. Both were the inheritors of a civilisation stretching back through the centuries to beyond the time of the Norman Conquest of England. The Senegalese empire of Songhay, which fell to the French, was the oldest and perhaps the more chaotic in its decline. The Fulani empire of Hansaland still possessed a nominal head. However, in both the actual power had been seized by a number of new adventurers, raiding right and left among their sedentary and civilised pagan subjects, and the whole country was being laid waste and going out of cultivation.

Since Barth's time in 1857 the population of Hansaland alone had been reduced to about one quarter of what it was then. The Fulani rulers, a race of remote white origin, had established themselves in a wave of postulating zeal a century before. Hitherto they had been pastoralists, partly indigenous to Hansaland but mainly coming from Songhay to the west. They had proved their worth under British administration.

At the time of partition the British government, faced with a series of small colonial wars, handed over the administration of Nigeria to the Royal Niger Company, under a Royal Charter. For the first twelve years the Company found itself fully occupied with the coastal provinces of Lagos and Southern Nigeria, a country of inhabited, dense tropical jungle. The Company naturally found its hands full.

France, meantime, had been more energetic in her sphere. Pushing up through the open and more or less civilised uplands of Senegal both westward from Dakar and southward from Algeria she had reached a point, where, if Britain did not wish to lose Northern Nigeria it had to do something to occupy it. In 1897 Captain (now Lord) Lugard[10] was sent by the Company to make a series of treaties voluntarily as the lesser choice of two evils, since the French were already threatening the northern borders while the British were still far away. In accepting British sovereignty, the Emirs undertook to open up the country to trade and refrain from slave trading, though the latter undertaking was made tongue-in-cheek.

Sooner or later a trial of strength was bound to come. The ink of the treaties was barely dry and the spark came with the two southernmost emirates of Nupe and Ilorin. These two Emirs in alliance started their annual raids for slaves into the Yoruba country, an industrious and hard-working community under the Niger Company's protection. The Emir of Nupe was the ringleader. Protests proving of no avail, a small force of Hansa Constabulary, all that was available, under Major Arnold, 3rd Hussars, and accompanied by Sir George Goldie, the High Commissioner, was despatched in January 1897 to bring the Emirs to order. The two Emirs combined were known to be able to put 30,000 armed men into the field, one third of them cavalry armed with muskets and a proportion of modern rifles. To meet this potential army the little expeditionary force mustered barely 600 men with 32 white officers and NCOs, a battery of 7pdr screw guns and a battery of Maxims. The Constabulary was armed with Sniders. White's success would have far-reaching results; defeat meant not only annihilation, but the practical ruin of the ten years of the Company's administration.

The little force started from Lokoja at the confluence of the Benue and the Niger Rivers on 6 January and fortune favoured the brave from the outset. Half the Nupe army had crossed to the west of the Niger on a raiding expedition and the armed launches of the Company were able to prevent the recrossing to defend the capital. The Emir of Ilorin decided, in the meantime, to wait for results. This had the effect of considerably reducing the odds.

---

[10] Later Lord Frederick John Dealtry Lugard, 1st Baron Lugard GCMG CB DSO PC. Born 22 January 1858, died 11 April 1945.

On 15 January the advanced guard drove in the enemy scouts, some 15 miles from Bida, the Nupe capital. On starting the march the next day, touch was at once regained with the enemy. The column marched in square and on topping a ridge 5 miles from Bida found themselves in sight of the walled capital and the whole available Bida army. The Bida cavalry made repeated charges but, after being turned back by the Maxims, they made way for their infantry sharpshooters. A lucky shell killed the War Chief, resulting in the contingents of two tributary towns of Lapai and Argai deserting during the night.

On 27 Sunday the whole force in square advanced to within two miles of the town. Again, the Fulani cavalry surrounded the square and made repeated charges but each time were repulsed without breaking the square. The Maxims and guns kept firing into the infantry masses, just outside the wall of the town. Once again, a lucky shell wounded the Emir and killed all his staff. A general retirement then took pace. The guns bombarded the town and two companies advanced to fire rockets onto the thatches of the buildings. This was too much for the warriors, who abandoned the town and took to the open plains, being heavily shelled till they got out of range. The battle was won and the Emir eventually surrendered to become the vassal of the Company. It is estimated that some eight thousand Fulanis took part in the battle.

As time went on occasional expeditions were necessary to bring a recalcitrant chief to hand: some were simply military and police promenades, others were of a more serious character, and it is this type of work that the mounted infantry fulfilled their natural functions. The account which follows is of mopping-up operations against the two truant towns of Argai and Lapai in 1898.

In spite of the easy victory Nupe was regarded on both sides as simply the first round in the coming struggle for mastery of Hansa land. The Niger Company did not possess the force to garrison even Nupe and withdrew after imposing terms on the Emir. An internal force had to be raised, known as the West African Frontier force, and in the following year, 1898, the Royal Niger Company surrendered their charter to the Crown as far as Northern Nigeria was concerned. The Crown finally took over the whole of Nigeria in January 1900.

The West African Frontier force, as its name implies, was originally raised for the Frontier Islamic Emirates, though afterwards it absorbed all local forces in British West African colonies. The original force, which later became the Northern Nigerian Regiment, consisted of two infantry battalions, the 1st and

2nd, each 1,200 strong; one mounted infantry battalion, the 3rd, 700 strong (raised by Captain (afterwards Lieutenant Geneneral Sir Thomas) Cubitt) and three batteries of Mountain Artillery. The 3rd mounted infantry battalion of the Northern Nigeria Regiment had the distinction of being the only mounted infantry battalion in the British Army to be raised and recruited as such, instead of being drawn from detachments from other infantry units. It also had the unique distinction of being shown in the monthly Army List as a recognised mounted infantry unit for over a quarter of a century. If, in the later years of its service, it was shifted to perform the duties of mounted police, in the first five years of its existence it took part in some very gallant fighting.

Trouble broke out shortly after the Crown had finally taken over from the Royal Niger Company in 1900. Half the infantry of the WAFF had been despatched to take part in the campaign in Ashanti and the Emir of Nupe in confederation with his northern neighbour of Kontagora, took the opportunity to try to eject the British from the country. There followed a series of violent skirmishes with the Fulani cavalry, who were decisively defeated with considerable loss. In all these the mounted infantry played the part of cavalry to the small forces engaged. The country was generally open, not unlike that of central India, and well adapted to mounted work. The WAFF, though almost invariably greatly outnumbered, were now better armed as the Martini Henry carbine had superseded the Snider.

At the end of December, the return of the troops from Ashanti enabled an expedition to be formed to finally bring the Emirs Nupe and Kontagora to book. Both towns were captured, their Emirs driven out and new Emirs installed on nomination by the Native Council of such states in accordance with ancient precedent. There was, however, one great change as confirmation of the Office of Emir came now from the Sultan of Sokoto – hitherto the supreme authority.

In September 1901 owing to the hostile attitude of the Emir of Yola an expedition had to be sent under Colonel Morland and, after obstinate fighting, the town was taken and the Emir, preferring exile, took flight and a new Emir was installed. A similar state of affairs occurred in a further expedition under Colonel Morland in February 1902 with the same result. All these fights followed the same course as that set by Bida in 1897, namely skirmishing by the Fulani cavalry in which British mounted infantry were fully employed and then a set battle in front of the town with charges by the Fulani cavalry. After repelling this the British infantry advanced under the cover of guns and a final breach of the town walls if the gates were not thrown open. The WAFF was, of course, by this time a more formidable and compact force than the gallant Hansa Constabulary of five years back. In all these expeditions there were clashes with turbulent tribes living in blocks of hill and forest country but these from the nature of the ground were the affairs of the infantry.

So far, the Sultan of Sokoto, head of the whole Fulani Confederacy, in spite of his treaties, once he found himself in real touch with the British, had refused any sort of cooperation. He had definitely refused to service his right as suzerain to nominate successors to the deposed Emirs and finally, in May 1902, he wrote a letter to the High Commissioner, as follows:

'From us to you. I do not consent that any one from you should ever dwell with us. I will never agree with you. I will have nothing ever to do with you. Between us and you there are no dealings except as between Mussulmans and Unbelievers ('Kafiri') War, as God Almighty has enjoined on us. There is no power or strength save in God on high. This with salutations.'

Received about May 1902. Source: *Colonial Reports – Annual, No. 409, Northern Nigeria Report for 1902.*

Kano, the capital of Hansaland, was in full accord with Sokoto. Kano town was deemed impregnable by the natives. In November 1902 the Emir of Kano marched against the British at Zaria but turned back on hearing of the death of the Sultan of Sokoto. War was, however, inevitable.

In January 1903 a small expeditionary force consisting of 24 British officers, 12 British NCOs and 722 rank and file with the four guns and four Maxims were assembled at Zaria and left for Kano under Colonel (later Lieutenant General Sir Thomas) Morland on 29 January. Eight miles after crossing the Kano frontier opposition was encountered from a small walled town and the gate was blown out by gun fire. The garrison therefore surrendered, and no further opposition was encountered on the road to Kano. The assault of Kano, if properly held, might have formed a formidable proposition. The town was surrounded by a wall 11 miles in circumference. The wall was 40ft thick at the base and from 30-50ft high. There was a double outer ditch, very deep and full of live thorns. The eleven gates were defended by enfilade fire from the walls. Fortunately, there were practically no defenders – a small breach was made in the wall and the defenders fled as soon as the heads of the storming party appeared. A considerable loss was inflicted on the garrison outside the walls as they attempted to retreat. On entering the town, it was ascertained that the Emir had left for Sokoto a month previously, taking the main body of his army with him and leaving the defence of the town to two slaves.

Kano was entered on 3 February and Brigadier General GV Kemball the IG, who arrived shortly after, took over command of the force. The Emir, meanwhile, was on his way back to Kano, when he heard that the town had surrendered. Thereafter a division of opinion arose in his camp. His brother, known as Wombai, maintained that further fighting was useless, and that the Emir should come to terms with the British and accordingly drew off half the force. The Emir then placed the loyal portion of the forces under his Wazir, whilst he himself fled in disguise toward the French frontier.

Kemball moved out to meet the returning Kano army sending a small force of mounted infantry, less than 150 strong, ahead to reconnoitre a day's march ahead of the main body. This small force under Captain W. Wright[11], with Captain Porter and Lieutenant Wells under him, came in contact with the enemy on 25 February and found they were up against the advanced guard of the Kano force of some 2,000 horsemen. Finding something of what they were up against the mounted infantry, under Wright, hastily formed a zariba out of thorn trees and bushes and repelled attack after attack for the next four hours, though surrounded on all sides, and until relief could come by the arrival of the main body next day. As soon as this was known, by enemy movements, to be within measurable distance Captain Porter led a daring charge. For this brilliant little action Captain Wright received the VC and Captain Porter the DSO, Lieutenant Wells receiving a Mention in Dispatches.

On 4 March Wombai, with his portion of the army, made a formal surrender and undertaking to deliver up all firearms and bows and arrows into British hands. Some 2,500 horsemen and 5,000 foot filed through the specified gate at Kano for this purpose. Kemball now moved on Sokoto, 200 miles to the northwest, securing reinforcements en route from the Boundary Commission at Argungu. A decisive action was fought under Kemball, on 15 March, outside

---

[11] Later **Brigadier-General Wallace Duffield Wright, VC, CB, CMG, DSO.** born 20 September 1875, died 25 March 1953.

Sokoto, when 1,500 horsemen and 3,000 foot were put to flight and in another severe action at Burmi, a few days later. Sokoto was entered on 10 March – the Emir, like his confrere of Kano, having previously fled. The trial of strength had come and gone and the Fulani Emirates henceforward acknowledged Great Britain as suzerain.

In this brilliant little campaign, which had lasted under two months, two Fulani armies had been brought to task, whilst the total British force available, inclusive of normal garrisons, never exceeded more than 1,500 men in the field, including mounted infantry. Though it is difficult, after the long interval, to get details, the many gallant deeds performed are partly asserted just by the list of honours won. Besides the VC awarded to Captain W. Wright and DSO to Captain Porter; Major C. Barlow (Essex Regiment), Captain S. Christy (20th Hussars) and Captain C. Maud (Somerset Light Infantry), all serving with the mounted infantry, were each awarded the DSO for their services in the operations with the Kano expeditionary force. Major Marsh (Royal West Kent Regiment) in the MI was killed and Major Barlow and Captain Christy wounded.

In the following few years there was still a lot of fighting to be done, chiefly against local natives, but the operations were more in the nature of police work than regular military operations.

Chapter 5

Mounted infantry in the South African war

The career of the mounted infantry as a recognised branch of the regular army, from the first establishment of the Aldershot MI School, expansion to a strength of 16,000 regular mounted infantry in South Africa and its gradual extinction by the eve of the Great War of 1914-1918, maybe, not inaptly, compared to that of a Beacon bonfire. It takes time after the first ignition for the fire to get any grip and then suddenly bursts into flames and, after an all too brief blaze, then sinks into glowing embers, which eventually spell extinction. The draught of the South African war provided the blaze.

In attempting to compile a history of the mounted infantry in South Africa the historian is met with certain opportunities which do not occur in any other branch of the service. Extension of the regular mounted infantry branch from its original two battalions to 16,000 men does constitute a rather unique phase in the history of the British Army.

Immediately after the war a Commission of Enquiry was held as to the future of the force. Strangely enough little credence of the work done by the mounted infantry was taken and the finding was unfavourable and not entirely free from bias. This had an unfortunate repercussion on the War Office since it apparently decided that the record was not worth retaining and consigned the regimental diaries to the wastepaper basket. There is therefore no official account of the various battalions that were raised.

The battalions themselves were subdivided into sections and companies from different infantry battalions. Private diaries that still exist record the narrator's experience from the company rather than the regimental point of view. These companies often served on detachment and could be transferred to another mounted infantry battalion as circumstances dictated.

The other reason for lack of records is the nature of the war in South Africa itself. With the railway and the only three towns in the country in

British hands there was nothing in particular for the Boers to defend. The two capitals, Bloemfontein and Pretoria, were small provincial towns, which owed such prestige as they enjoyed to the fact that they were the seat of government - a government which legislated for the simple needs of a settled and pastoral community. Johannesburg, centre of industry, was alien in outlook and the cause of half the trouble.

The Boer forces, split into their territorial units and commandos, carried on the struggle as opportunity offered. This dispersion of the Boer forces as a cohesive whole necessitated an almost equally wide dispersal of the British forces. Henceforward the history of the war was that of scattered columns, sometimes acting in conjunction but often more or less separately. The Boer commando occasionally combined for some specific object, like de Wet's raid on Cape Colony, but more often confined themselves to keeping out of harm's way and harassing the British whenever an opportunity offered.

The Times and official historians of the war dealt with the phases of the war than the doings of any particular column. The columns themselves were often diverted from one scene of action to another, over a country as big as France.

The only connected history of the mounted infantry is contained in their association with French's cavalry in Lord Roberts' advance to Pretoria. During this time the first eight mounted infantry regiments served as a division; though the two brigades in which they were formed were by no means always brigaded together.

An effort will now be made to follow their adventures during this period, which started from February to June 1900 inclusive.

The calvary, at the start, regarded their mounted infantry comrades with good-natured tolerance. They were useful for rear-guard or as escort to the baggage convoy, also to hold and cover a drift or a *donga*. They might even, on occasion, be asked to brush inconvenient forces off a *kopje*, but they fell far short of cavalry conceptions of mounted warfare. In the end the cavalry found themselves compelled to adopt the same despised mounted rifles tactics, but not yet a while.

Before going further, however, it is as well to glance at the position of the British Army just before the outbreak of war, for it explains many of the initial mistakes and failures at the commencement of the war. The army was undergoing a leisurely transition. Except for colonial wars, Britain had been at peace for many years. The outlook of the higher command was acutely dominated by the lessons and experience of the Franco-German war of 1870. Only seven years before the South African war broke out the infantry at the Curragh, though rearmed with the new bolt-action Lee Metford rifle, were still firing their annual course of musketry with the heavier-calibre lever action Martini Henry officially retained for the purpose! Route marches and field days were still being carried out in revised order with pack. Extended order was at two paces interval, anything wider being considered unmanageable. With no experience of high explosives, entrenchments were still those of the Crimea with a wide trench for free movement and a high parapet to fire over. The numerous trenches of the Boers with no parapet were a revelation.

The calvary were trained to rely almost entirely on the *arme blanche* and shock tactics, and they regarded the carbine as a purely subsidiary weapon. The lance, especially in pursuit, is a terribly demoralising weapon and its effect at Elandslaagte did make itself felt in the earlier battles of the Western Free State. However, it is useless against modern infantry and machine guns who stand their ground.

Light artillery fire was credited with far more effective power than later was found to be the case. Such a thing as a modern methodical artillery carriage was unknown at that time. Initially, the horrendous power of modern rifle fire was not understood. In March 1899 a senior lieutenant[12], just home from frontier wars in India, was sitting his examination for captain. Among the questions asked was the candidate's views on the following situation: 'A battery in the open comes into action against an infantry company (100 men) under cover at 1200yds range.' His laconic answer of, 'The battery would be wiped out,' sustained by examples from his own experience, so surprised the War Office that he was sent for and officially interviewed!

Now, the Boers were, of course, a *levee en masse* of the fit male population. They were untrained for warfare and at the start possessed little discipline. They were only half-inclined to accept the advice of their foreign military advisors. They were, on the other hand, almost all countrymen, with a natural eye for country and accustomed to ride and shoot from boyhood. They went to battle at the start in the same spirit as they went to quarterly *nachtmahl*

[12] The author - G. Standish Craufurd.

(communion). That is to say, in their ox wagons accompanied by their wives and children, leaving the older men and invalids with their native labour to mind the farms. This naturally hampered their movements. From overwhelming confidence at the start, they moved to despondency as the might of Britain unfolded. Then, as a better understanding came, they settled down to the fine, dour guerrilla fighter that they afterwards found themselves to be. Individually there was nothing but language to separate the two races.

When Lord Roberts arrived at Cape Town, on 10 January 1900, the situation was not very promising. In the east Ladysmith was besieged and Buller's relieving army was held up at Talana. On the central railways French and Gatacre were facing an invasion of Cape Colony. In the west Methuen had suffered defeat at Magesfontein and Kimberley was already invested. Something had to be done urgently to restore British prestige.

Meantime, reinforcements were flowing into the country and local mounted corps were being formed. The 1st and 2nd Mounted Infantry Regiments had been formed from the MI School at Aldershot and sailed with the Cavalry Division.

Every infantry battalion in South Africa had been told to form a mounted infantry company, with the exception of the Ladysmith Garrison who were, of course, cut off. From these companies or sections another six mounted infantry battalions were formed. Some of these companies and battalions were only just organised and had had no mounted infantry training when Lord Roberts' advance began. One or two had been fighting with French at Colesberg and were thus settled. Others had even lacked equipment other than saddlery and their mounts when they marched off into the unknown and action.

These eight MI battalions were formed into two brigades forming an MI division attached to French's Cavalry Division, which consisted of three cavalry brigades and six batteries of RHA. The 1st Mounted Infantry Brigade under Colonel Hannay consisted of the 1st, 3rd, 5th and 7th MI Battalions, and the 2nd MI Brigade under Colonel Ridley of the 2nd, 4th, 6th and 8th MI Battalions.

There were two lines of obvious approach into the Free State: the direct one via Colesberg and Bloemfontein and the westerly one via the Orange River and Kimberley. Roberts decided to take a middle course, taking the westerly route to just short of the Modder River and then, turning east along the Modder River, attacked Bloemfontein from the west. Kimberley was to be relieved en route. The plan involved a good deal of preliminary counter feint and camouflage to be successful.

The concentration of the 6th, 7th and 9th Infantry Divisions as well as the cavalry and MI took place in the neighbourhood of the railway some 20 miles south of the Modder River and the move forward began on 11 February.

To mark the great concentration and further perplex the enemy certain demonstrations and side shows were staged during the first week of February in which the cavalry were chiefly concerned. These marches in a dry and thirsty land on top of much hard work at Colesberg took it out of the cavalry horses. On 12 February the cavalry, covering the vicinity of the railway, marched west and secured the crossing of the Riet River, the first obstacle on the path.

On 13 February the Cavalry Division, after a 30-mile march, had secured the crossings over the Modder River at points 12 and 17 miles upstream from Magersfontein and the way was open to Kimberley. In manoeuvring for the drifts much more than 30 miles was covered, and the horses needed rest. It was also essential that the infantry should come up to hold the drifts whilst the 25-mile dash to Kimberley was made and in spite of the urgency a 24-hour rest was inevitable. It was here the mounted infantry caught up and joined the cavalry. It was to be an association that lasted four months from February to June 1900 inclusive and during that time both were to learn much from each other.

The country they were now operating in, the Western Free State, was flat and sundry. It was African midsummer and the midday sun heat was intense in those arid plains. Except for the rivers, water was only to be found in rare intervals at occasional dams. Grazing for horses and oxen was poor, burnt up and the marches proved very hard on the animals. The two rivers, Riet and Modder, had cut deep beds in the sandy soil, the steep riverbanks covered with mimosa plants. The only access to the river was by the various drifts or crossings cut by the Boers for their wagons and for watering their animals.

The cavalry baggage only arrived at 4.00 pm on 13 February and Kelly-Kenny's 6th Infantry Division did not arrive till just before dawn on the 14th. Regimental transport was drawn by teams of mules but the supply train from which they were refitted was drawn by slow-moving oxen. In all these earlier marches it was a constant question of lightening the weight on the horse or lightening the forage ration to enable the baggage to keep pace with the cavalry. Hence the poor condition of the horses.

At 9.30 am in the morning of the 15th, French started on his ride to Kimberley. Cronje was still at Magesfontein 12 miles to the east and there was a chance that the march would be unaffected. Roberts' orders were to relieve Kimberley that day if it cost French half his force to do so.

No sooner were the scouts well over Klip Drift than they came under fire. The Boers were holding a semi-circular position on higher ground some three miles from the river. This position was lightly held in front but well covered by fire from either flank.

There followed what was possibly the last, great, spectacular cavalry charge in history. Six thousand horsemen, in open order the Lancers leading, many lines deep, rode in a vast high cloud of dust straight into a five-mile line of riflemen! On they rode forward and upward until they disappeared in a cloud of dust in full view of the infantry and MI still at the bivouac by the riverbank. The Boers fired wildly into the charging horses, then seeing their fire was not taking effect and horse and steel bearing down on them they were overcome by panic, sprang to their horses and fled – some even to Cronje's *laager* 12 miles away. The way to Kimberley was open.

A halt was made at Abonsdam, half way to water and feed, and the siege of Kimberley was formally raised at 6.00 pm. It was a great start for the four months' association of the regular mounted infantry and the Cavalry Division.

French's breakthrough had placed Cronje in a critical position by cutting off any hope of retreat to the north and the infantry advancing in French's wake would cut off his retreat to Bloemfontein and the east if he did not move at once.

Bloemfontein lay to the south of the Modder River; Klip Drift and Klip Kraal further upstream were already occupied by the British. Paardeberg was the nearest practical drift available and that might be occupied any hour. He decided on Koodoosberg Drift, 45 miles away. Cronje lost no time in moving. He moved that afternoon in French's tracks still fresh on the sandy soil, in full view of the mounted infantry battalion French had dropped at Abonsdam, five miles to the south, to safeguard that small water supply for the oncoming infantry.

Both sides glared at each other from a respectful distance. Cronje did not wish to engage in a small battle in the heat at that moment and the mounted infantry were told not to interfere. Moving by forced march he had covered the 30 miles to take him round Klip Drift by the evening of the 16th! He bivouacked for the night at the little drift of Vendutie two miles east of Paardeberg, which was unsuitable for wagons and only suited for watering.

Cronje felt secure for the moment. French, he knew, was 35 miles away north of Kimberley and he had 15 miles' start on the nearest infantry, which would have given him time to cross the drift at Koodoosberg and get clear away. He had outwitted the Rooineks this time.

After his forced march it was necessary to graze his horses for a few hours in the early morning and by 10.00 am on the 17th he had just inspanned prior to resuming his march when his scouts on the low hills to the north came galloping in with the horrifying news that a British mounted force had appeared out of the blue. Who they were and where they came from no one could guess. At that moment French's guns came into action. The surprise was complete and their effect instantaneous. The convoy halted dead in its tracks, the oxen were outspanned and driven to the riverbed along with the women and children.

French, in his dash for Kimberley on the 15th, had left the mounted infantry with his train, except for the detachment at Abonsdam and the escort to the guns. Early on the 16th the MI at Abonsdam got through the news of Cronje's passing. At the same time the infantry outposts at Klip Drift reported the movement of Conje's *laager* and the MI got in touch with his rear-guard.

Kitchener, who was present, got things moving at once and urgently sent a message to French to co-operate. French, with tired horses and a long march before him, could only take his fittest mounts. At 3.00 am of the 17th a small force under Broadwood, 1,000 strong, with a brigade of Horse Artillery left Kimberley to be later joined by French with another 200 men. Six hours later, at 9.00 am, patrols reported clouds of dust in the direction of the Modder. The dust was behind two groups of *kopje*s, the western end of which appeared to be a mile or two beyond the dust. French made a dash for the eastern side. It proved to be the Koedoesrand overlooking the left of that area and Cronje's *laager*, three miles long, lay inspanned obliquely to the west. A few prisoners captured on those hills confirmed the situation.

French had certainly startled the Boer convoy but he was only 1,200 strong facing 4,000 Boers. All day long French lay on that semi-circle of hills watching the Boer *laager* in the plain below and anxiously scanning the horizon for the oncoming infantry. For in the distance dust could be seen but that was all.

Sniping went on with Boers inside and those outside on the plain, which died down at dark. Fortunately, the Boer *laager* was in complete confusion at the unexpected turn of events. French's small column bivouacked that night in the positions they had occupied all day.

The 6th Infantry Division went into bivouac on the morning of the 15th at Klip Drift. It moved across the drift at 4.30 am on the morning of the 16th with Hannay's MI as advanced guard. The men of the 6th, which were the vanguard, had hardly left camp when they ran into Boer wagons and learnt from prisoners that Conje had just crossed their front. As the light improved, they could see clouds of dust rising beyond the low ridge which had formed the left of the Boer position when French made his charge the day before.

General Knox, with the main body of the leading brigade, also saw the dust and realised its impact. He immediately swung his brigade to the right and arranged with Hannay for a concerted attack. The first range of *kopje*s was reached by the 6th Mounted Infantry who saw the mass of vehicles passing the edge of another line of *kopje*s four miles further on. They were the first of the army to see a great Boer column in flight.

The Boer rear-guard was well handled and withdrew to the second ridge known as Drieput. Here the Boers held a strong position with other hills joining up to the Modder River some seven miles away. This position the Boers held till nightfall in spite of Knox's attack, which was pressed on to close range but could not sway by assault.

Starting on the extreme left and engaged with the Boer rear-guard in the first action when they fell back to the second position, the 6th Mounted Infantry were sent to the extreme right, some 6 or 7 miles wide. It was later sent to the south side of the river to support an infantry approach to Drieput Drift. The experience was an arduous one entailing, as it did, a running fight all day.

Cronje meanwhile had made a 14-mile trek under cover of his rear-guard and after halting at Vendutie Drift to rest and graze his cattle had just inspanned for his forward march when French came upon the scene. He had confidently believed that he had shaken the infantry off.

After a hard day's fighting Hannay's MI had only managed to cross Klip Kraal Drift three miles further on. By 10.00 am on the morning of the 17th the MI again got in touch with Conje's rear-guard and again a running fight resumed all day. Towards evening they rejoined the 6th Division south of the river and marched on Paardeberg Drift. The Division lost its way in the dark and at midnight bivouacked two miles from the river.

It was still possible for Cronje to break out that night, in the hope of rejoining some of the commandos outside, but it meant abandoning many of his wagons and weaker teams.

The Boer *laager* lay in a shallow cup of encircling hills which the river cut in two. To the north the river of the cup was well defined with its edges framed by the conspicuous *kopje*s of Paardeberg and Koodoosrand 10 miles apart and an arc of lower hummocks and hills (occupied by French) in between. To the south a belt of green – the tops of the trees in the riverbed showed where the river ran. On the south side of the river the rim was more broken and the circle of hills further away but in the centre, 4,000yds from the *laager* and four miles from the outer rim, was a prominent hill, later known as Kitchener's Kop, which was occupied by de Wet at the time.

Kitchener determined, quite rightly, to carry the *laager* by assault that day. At that stage of the campaign a successful assault, the first by night, might have had far-reaching effect.

The plan required simultaneous assault, which in turn demanded time and a carefully coordinated plan. Unfortunately, zero hour in those days was an unknown quantity. Kitchener took the more sporting view of a hunter having a pack of hounds on a quarry at bay.

By dawn the 6th and 9th Divisions were coming up into position with Hannay's MI in the vanguard.

Kitchener's plan for attack was as follows: Kelly-Kenny with the 6th Division was to attack frontally south of the river. The MI supported by half of Stephenson's brigade was to push onto Vanderberg Drift, two miles above the *laager*, which was already lightly held by a cavalry piquet, and attack downstream from the east, whilst the 9th Division was to cross the river at Paardeberg and attack upstream from the west. The presence of de Wet on Kitcheners' Kop was overlooked as of minor importance. The artillery was to co-operate. None of the attacking units had any knowledge of the role to be played by their neighbours. The attack was to start at 8.00 am.

The forward troops of the infantry were only just arriving on the scene after a long march and deployment took time. The hounds ran joyfully into the kill independently as they arrived on the scene and were severely held up in turn.

Taking the MI, with whom we are chiefly concerned, the 7th MI on detachment began to work up through the south of the riverbank from Paardeberg Drift below the *laager.* They were getting on nicely when they were reported as masking the infantry fire. The 6th MI on advanced guard performed a similar move from Vanderberg Drift above the *laager* and spent most of the day in the riverbed.

Hannay, his force of 1,500 men by now reduced by various detachments, crossed the Drift and started to attack and once within effective range, got hung up. De Wet on his *kopje* lay low till the 6th Division were committed to their attack. He then started dropping shells from two guns on the *kopje* indiscriminately on any good target he saw. If he did not cause much damage, he succeeded in causing considerable confusion.

Stephenson's two battalions, not realising they were to support Hannay, turned wide to chase de Wet off his *kopje.* De Wet thereupon cleared off, hanging about *kopje*s in the rear and sending messages in helio to the beleaguered *laager* - most of them quotations from scripture. To some of them Kitchener replied, and the correspondence duly noted was captured by the 6th MI a year later. Stephenson's battalions were directed back to support Hannay's advance, arriving a bit late.

It was now two o'clock and the British force lay extended and unable to advance further on their sides of the *laager* some 700yds away. Hannay reported to Kitchener that he could get no further but Kitchener was getting desperate for a final effort to rush the *laager* before dark. He urged Hannay to do something, anything, even to try to rush the *laager* at a gallop! He omitted to say that he was going to call on everyone for a final effort. Orders were orders. Hannay argued no further. He collected fifty officers and men for what he knew was a forlorn hope and with these he charged accordingly. The charge advanced the line another 200yds to a fold in the ground where they were to remain till dark. Hannay and several of his officers and men were killed.

Two hours later 18th Brigade under Stephenson aided by a group of MI made a final effort and achieved nothing. Darkness fell and the ten-day siege commenced. A little more staff work and the day might have provided a brilliant success. For the remainder of the siege the MI reinforced the cavalry on outpost duty.

De Wet had first come into prominence three days earlier on 15 February when he attacked a convoy parked at the Reit River. The convoy was defended all day by a company of Gordon Highlanders and some MI against heavy odds. General Tucker, with the 7th Division, was on his way to relieve it when he was diverted by Lord Roberts owing to news of Cronje's move. The convoy was consequently abandoned. The loss particularly affected the MI as it contained blankets, breeches, bandoliers, forage, provisions and other equipment for them as well as putting horses and men on half rations for the whole force until Bloemfontein. It was the ragged appearance of the mounted infantry, in consequence, during the march to Bloemfontein which first inspired Rudyard Kipling's poem of *Ikona.*

During the siege Boer commandos began gathering in some strength in the east and south. Hitherto French had only occupied the low hills between Paardeberg and Koodoosrand with his left flank drawn in towards Vanderberg Drift. It now became necessary to seize the Koosdoosrand, which commanded the Drift of that name. The mounted infantry therefore took over from French leaving him free for further operations.

Koosdoosrand was then occupied on 21 February without any serious opposition and a gathering of some 2,500 Boers dispersed. From Koosdoosrand the rim of hills south of the river was also cleared the next day. This freed Kitchener's Kop, which had remained a thorn in the side of the British during the first four days.

On the whole the ten-day siege proved a period of rest and recuperation for the mounted infantry. During the period and the weeks' halt succeeding the surrender, the mounted infantry, including colonial contingents, were reorganised into four brigades under Ridley, Alderson, Le Gallais and Martyr respectively. These brigades sometimes worked with the Cavalry Division; at other times they were detailed to work as mounted troops covering the infantry.

Driven from the neighbourhood of Paardeberg the Boers took up a position on a line of hills 12 miles further on known as Poplar Grove. It was in full view of Koodoosrand and reconnaissance showed that they were putting it into a state of defence. The position, a very strong one, astride the river and stretching in all some 10 miles, was designed to be held by 12,000 Boers with guns and President Kruger came down to harangue them.

North of the river was a big, flat-topped hill known as Leeuw Kop. South of the river was an area more formidable than Table Mountain and beyond this, again connected up with it, ran a chain of seven *kopje*s running southwest.

The disposition for attack was as follows. The cavalry with Alderson and Ridley's MI were to march at night and by a wide flanking movement to the south, out of sight of the Boers, were to get behind them and cut off their retreat. The 6th Division with Martyr's MI were to follow in the cavalry wake to a point southeast of the seven *kopjes*, then turning inward to take the position in the flank. The 7th Division, with half of Le Gallais MI Brigade, were to make a frontal attack following the line of the river. The 9th Division were to move north of the river with the other half of Le Gallais MI guarding their left flank.

The attack developed into a spectacular parade. The cavalry were late and instead of getting behind the position mainly arrived in time to turn the left flank. The infantry advance was slow and methodical, not to say cautious. The battalions advanced in column of companies at 20 paces interval, a first attempt at artillery formation, and covered the whole plain. The mounted infantry rode wider on the flanks. To the Boers on the grandstand this steady and extended advance of 25,000 men, supported by guns in the rear, reinforced by the more distant baggage train, must have appeared like a gigantic review on Laffan's Plain and was doubtless most impressive even if red coats and bands were not about. It should have ended of course in a *feu de joie*. As a matter of fact, the attention of the spectators was somewhat diverted by the disconcerting appearance of the cavalry on their left rear. There being no loudspeakers, stout hearted old Kruger's fiery eloquence failed to direct the attention of the audience to the show in front. They panicked and fled, leaving only a stout rear-guard to see the performance out. These, however, did manage to keep up the interest till late in the afternoon.

The previous three weeks' continuous hard work and half rations of corn prevented the cavalry from making any attempt at pursuit. A two-day halt followed and on 10 March Lord Roberts resumed his forward march.

The Boers, including the foreign contingent, were rumoured to be in some strength at a group of *kopjes* 13 miles ahead on the south bank of the river, whilst Presidents Steyn and Kruger were known to be present. But, after the fiasco of Poplar Grove, it was doubtful if any real resistance would be offered.

Lord Roberts moved his force in three columns. In the centre moved the Commander-in-Chief with the 9th Division, the Guards Brigade, the 2nd Cavalry Brigade and Le Gallais and Martyr's MI. On the right the 7th Division,

3rd Cavalry Brigade and Ridley's mounted infantry. On the left the 6th Division, 1st Cavalry Brigade and Alderson's mounted infantry.

The right and centre met without opposition, but the 6th Division and 1st Cavalry Brigade came in for a pretty stiff fight at the aforesaid *kopje*s. Arriving at the first *kopje*, Abraham's *kraal*, the cavalry attacked the hill without great difficulty and the Boers fell back to the second position 7 miles further on at Dreifontein. Here they put up a most determined resistance. The mounted troops were held up and could not advance so fell back for the infantry to attack. The fight went on for some time, from ridge to ridge, until finally, towards dark, a most gallant assault was made by the 18th Infantry Brigade supported by the 13th.

Broadwood's cavalry had, in the meantime, been sent from the centre column and arrived in time to turn the Boer flank. This was too much for the Boers who fled in what de Wet himself described in his *Three Year War* as 'a disorderly crowd of terrified men flying blindly'. They left over 100 dead on the field to be buried by the British. The British, on their part, had one officer killed, sixteen wounded, fifty-seven rank and file killed, 321 wounded and twenty-six missing. For the numbers engaged these were quite heavy casualties at this period of the war.

The infantry of the 9th Division arrived at 5.00 pm, just too late to take part in the fight after a 16-mile march. Bloemfontein was entered by the cavalry on the 13th without further opposition. It was just a month and a day since the main advance had started.

Judged by the standards of 1914-18 or even today [1939-1945] the fighting had not been severe. With the exception of the fight at Paardeberg the mounted infantry fighting had chiefly been confined to skirmishing with rear-guard.

On arrival at Bloemfontein the troops found themselves in new surroundings for, though they merge unpresumptuously, the high *veld* is a different country to the hot sandy plains which border the Kalahari Desert. The soil is darker and the rainfall more frequent. The country is covered with a coarse grass resembling wild oats. From time to time there are wide stretches of cultivation, kaffir corn, mealies, oats and vetch. Every farm has a pond of muddy water for its livestock and a series of small rivers seam the country from east to west. The far more frequent farms are surrounded with orchards of peaches and often, fig trees.

To the mounted infantry these amenities were to mean so much in the days ahead. Although looting was strictly forbidden, the occasional leg of mutton or chicken hanging from the saddle or dried apricots in the surplus nosebags showed that the local population were practising appeasement before Neville Chamberlain gave it an international significance.

After the arrival at Bloemfontein the infantry had some six weeks of recuperation and rest, whilst preparations were made for the further advance. Not so, however, for the unfortunate mounted troops. Remounts were received but they were started off on a trek shortly after arrival with little opportunity to get acclimatised. It is due to this that the tremendous wastage of horseflesh in the advance from Cape Colony to Pretoria must be attributed.

The first calls were for a flying column to the east under General Broadwood. Some 20 miles east of Bloemfontein there is an outcrop of broken country, stretching out as far as Drakensberg, dominated by two flat-topped hills or table mountains some 10 miles east and known respectively as *Thaba* and *Thaba Nchu* (hill of sheep). This district was to form the scene of much desultory fighting during the next month as columns of Boers moved up the Basutoland border from Colesberg and Cape Colony in front of British columns moving north to join in Lord Roberts' advance.

For the moment the district remained peaceful enough. The Modder River has its source in these hills near *Thaba Nchu* and there the water works for the Bloemfontein supply were situated at Sanna's Post. The first objective was to make good the water works.

Chapter 6

6th Mounted Infantry march to Bloemfontein

The arrival of Lord Roberts in South Africa, to take command of the Western offensive, saw considerable preparations for the immediate increase of the mounted forces including mounting Australian and other colonial forces as fast as remounts could be secured. Yeomanry were despatched from England, in addition to the eight mounted infantry regiments abstracted from the regular infantry. These eight early MI wore a special device in their caps formed of two black ribbons sewn in the form of an arrowhead with the centre triangle of coloured cloth resembling the letter 'A' with the apex filled in. That of the 6th MI was black sides with red apex. This device was discontinued for MI raised later in the campaign. The fact that these earlier MI were composed of complete regimental companies contributed by battalions in the field - strengthened perhaps by the special regimental badge - developed an early regimental spirit, which later regiments with composite companies and often split into detachments found more difficult to achieve.

The 6th Mounted Infantry was fortunate in that it continued from the date of its formation to the day peace was declared as a unit continuously engaged on column work in the different storm centres of the Transvaal, Free State and Cape Colony. It also owed a very great deal to its fine fighting record of the spirit installed into it from the start by Colonel De Lisle, its first regimental commander and later column commander, almost up to the close of the war. The 6th MI was composed of four regiment companies: Bedfordshire Regiment, Welsh Regiment, Wiltshire Regiment and Gordon Highlanders.

There were, in addition, two sections of the Essex Regiment which were first attached to 'D' Company but six weeks later were transferred to 'C' Company with whom they continued to serve as an additional wing under their own officers to the end of the war. Lieutenant Col De Lisle of the Durham Light Infantry was commanding officer with Captain Robert Fanshaw, Oxford Light Infantry, as adjutant.

If some of the later MI were to have a little preliminary training in mounted work before being thrown into the firing line, the original MI units, with the exception of the 1st and 2nd, had none. Half the men had some acquaintance with horses as grooms, transport, drivers etc, but the remainder did not even possess this qualification. All were, however, old soldiers with some acquaintance of South African warfare.

'D' the Gordon Company was home to volunteers from the 1st Battalion at Maple Leaf camp on 8 February and left the regiment 105 strong under Lieutenant J.U.M. Ingilby next day to join their new formation, which was forming at Orange River Station. Leaving their kilts behind they started off in their fatigue trews in lieu of riding breeches. Putties had to be roughly improvised on the spot by tearing blankets into strips for the purpose. All were in Highland shoes, no boots being available; Lieutenant Ingilby alone was the proud possessor of riding breeches, which he had managed to exchange with Colonel Forbes McBean for his kilt. On arrival at Orange River, they drew saddlery, a field forge and a heap of horseshoes, both the latter having to be immediately dumped on the *veld* for lack of transport. That same night, on 10 February, they rode out and off to meet the enemy. Adventure began leaving the camp; sufficient time was not allowed by the head of column for the companies to close up after crossing the bridge over the Orange. Once over the bridge on fresh Australian horses and unaccustomed Australian saddlery, many in the saddle for the first time, rode into the dusk at a gallop to catch up the column on its three-year adventure. Girths broke, men and saddles hung from their horses at every angle, but they arrived in their place in order march, still mounted to a man.

The regiment, together with the 4th MI and Kitchener's Horse, formed an MI brigade under Colonel Hannay, Argyll and Sutherland Highlanders, known at the time as the II Mounted Infantry Corps. At daybreak on 11 February 1900, they encountered the Boers in superior force at Ramah Springs near Ram Dam on the Orange Free State border. They immediately engaged and held the enemy till French's cavalry swept past in their great flank ride to Kimberley. The Gordon Highlanders' casualties in this, their first fight on horseback, were three killed and eight wounded. They came out of it with flying colours, for their new CO came up and complimented them with the remark that, 'If they could not ride, they at least knew how to fight.' Hannay's

Brigade halted at Ramah to join up with the 9th Division under Major General Sir H. Colville as mounted troops.

They were thus at the start of Robert's march for the relief of Kimberley and the invasion of the Orange Free State. The dust raised by these divisions on the move, the still air, the hot day and lack of water made the march a very fatiguing one alike to infantry and mounted men. There was only one well at the camp at Ram Dam for the whole division, which was besieged by thousands of thirsty men. A dead local had been thrown down the well by the retreating Boers and was, unfortunately, not discovered at the outset. Many were consequently violently sick as a result of quenching his thirst.

Roberts with the two other divisions, French with his cavalry and MI were ahead, and Hannay's brigade was sent to reinforce. A further move the following day brought the brigade to the Riet River, the banks of which presented a most variegated appearance as hundreds of heavy lumbering transport waggons negotiated its steep sides. The day's scrap, as advanced guards to the force, was to give the 6th Mounted Infantry their first lesson in mobility in action. Starting on the left flank they were withdrawn and, trotting round behind the infantry, took up their position in the firing line again on the extreme right. During this manoeuvre a shell landed under the belly of a horse in front of Lieutenant Ingilby. Both man and horse were enveloped in dust and when it cleared the man was on the ground holding the reins, both uninjured. The shell was, of course, a dud.

Hannay's brigade moved on again at 1 am and marched to within 4 miles of Jacobsdaal. It was whilst camped here on 14 February the company heard of the loss of the convoy of 200 waggons at Riet River, which had been so pluckily defended by 'B' Company 1st Gordon Highlanders under Captain K. Dingwall and a squadron of Kitchener's Horse. This convoy, after being defended all day by Captain Dingwall despite fire from four of de Wet's guns, was eventually abandoned by order at dusk since the necessary troops to extricate it were more urgently required in front. The loss, however, was a serious one to the Gordon Company as the convoy contained, among other things, all the breeches and other riding kit for their proper equipment. Blankets were also short and for the time being there was seldom sufficient extra clothing available at night. The whole force was on three quarter rations and the MI were worse off than the infantry in that respect (a position of affairs which later on in the campaign was to be completely reversed). As an offset to the loss of the convoy the news of the relief of Kimberly was received the same night.

Jacobsdaal, a small village of irregular corrugated iron and thatched houses, was entered on 15 February. Inside the church flying the Red Cross were a number of British and Boer wounded attended by a German Red Cross nursing staff. Soon the several stores were engaged in a roaring trade with the troops, which left the storekeepers vowing that they would not object to an invasion like this every day.

On the afternoon of 17 February the force was again on the move, now in a definitely easterly direction following Cronje's *laager* after the relief of Kimberly. The march was a heavy one in choking dust and bivouac on the Modder River was reached at 2 am. After a little over an hour's halt the startling rattle of musketry nearby awoke the tired men. Patrols had come in touch with the outposts of Cronje's *laager*. In the early hours of the dawning light, Hannay's Brigade rode parallel with the left bank of the river, the gunners meanwhile shelling the *laager* over their heads from a position away on the right. When well above the *laager*, 'D' Company of the 6th MI received orders to leave the horses on the left bank, and cross over on foot and attack the *laager* downstream from the other side. The men crossed about six at a time, linking arms as the river was swift and deep; but the company got across without a casualty. There were fairly thick trees on the other side and they were busily engaged driving off parties of Boers as they worked in extended order downstream keeping as far as possible in line with 'C' Company doing the same on the left bank. Shortly after this Lieutenant Ingilby commanding the company was hit and put out of action. Hannay, with the remainder of the brigade, had meanwhile crossed the river and was prolonging the line westward to complete the circle of attack.

By 9 am the position was roughly as follows. South of the river were the 6th Division upstream and the 7th Division downstream, holding the two fords, above and below the *laager* and the country between. North of the river lay the 19th Brigade of the 9th Division with the Highland Brigade south of the river and joining up with the 7th Division. While on the left of the 19th Brigade were Hannay's mounted brigade, later reinforced by French's cavalry from Kimberley, completing the circle to the north and linking up, respectively, with the left flank of the 19th Brigade and the right flank of the 6th Division.

With Cronje thus completely surrounded an attempt was made to carry the *laager* by assault, the line south of the river acting as a containing force while the line north carried out the attack. While the 19th Brigade advanced on foot, Hannay attempted a charge on horseback. The Boers were well entrenched and no time was given for an efficient preliminary bombardment. Both attacks were therefore doomed to failure. While the 19th Brigade advanced to within 400yds of the *laager* only to find a final charge impossible, Hannay's charge was held up shortly after starting. Hannay was killed, his Brigade Major, Captain Malcolm, Argyll and Sutherland Highlanders, wounded, and among other casualties the 6th MI lost Captain Waldy and Lieutenant Selous killed and Lieutenant Jebb wounded, all of the Bedfordshire Regiment. Nothing remained but to extricate the troops from an impossible position at dusk.

On the following day a twelve-hour armistice was granted at the request of the Boers to collect the wounded and bury the dead. Lieutenant Jebb, who had been taken prisoner, was handed over by the Boers during the armistice. It was now recognised that direct assault was impossible. Siege operations were undertaken by the infantry while the mounted troops were withdrawn to the outer circle as containing troops and escort to the guns.

To Colonel Martyr, who succeeded Colonel Hannay in command of the brigade, was assigned the small group of hills some three miles north of the river known as Kamelfontein as centre of his line. This was disposed of as follows: 4th MI on the left as far as the river and ford; 6th MI in the centre; Kitchener's Horse connecting up with the infantry on Paardeberg *kopje* from which the battle took its name. The farm on which the actual *laager* was situated was known as Stinkfontein (from Stinkwood trees in the riverbed which gave a pungent smell when burnt). This name, however, was too appropriate to the development of events to be allowed to go down in history.

On 23 February, Captain Robert Gordon of the Queensland Mounted Rifles (then attached and later gazetted to the regiment) was sent to take over command of the mounted infantry company. Captain Gordon had been attached to the regiment in the Tirah campaign of 1898, was a universal favourite and a first-rate leader for guerrilla warfare. Affectionately known as 'Boomerang' from the performances he used to give with that weapon to admiring Afridi Jirgas in Tirah. He continued to command the Gordon Company till wounded in Cape Colony a year later.

On 26 February the company formed escort to the guns shelling the northeast corner of the *laager* and came in for a good deal of long-range fire. During the evening, following a rumour that Cronje would try to break out, the MI lay out during the night at five-yard intervals with two companies forming an inlying picquet in reserve.

Early in the morning of 27 February, when the investing forces were waiting prepared for a final assault, the white flag was hoisted and Cronje with his commando surrendered.

Hard put as the whole investing force had been on three quarter rations, even at this early stage of their experiences as mounted soldiers, the MI had found opportunity to supplement their meat rations with slaughtered bullocks when foraging at outlying farms for grain for the horses. The horses themselves fared badly as grazing was very poor. The week after Cronje's surrender passed quietly in camp on the old picquet line, and some instruction in horse management and riding was given but the horses were in too poor condition to allow of much riding school.

On 7 March the Martyr's Brigade struck camp at 3 am and moved out in the dark to take their share of the action at Poplar Grove. The part assigned to Colonel Martyr's MI corps was to cover the left flank of the infantry advance. The brigade was disposed as follows: 6th MI on the extreme left of the army; 4th MI in the centre; and Kitchener's Horse on the right linking up with the 1st Gordon Highlanders.

In front of the 19th Infantry Brigade loomed the formidable position of Leewkop, held as the Boer right flank and dominating the country north of the river. South of the river the 7th Division moved on the Boer centre while French's cavalry with the remainder of the mounted infantry attempted to outflank the Boer left on Seven Kopjes supported by the 6th Division. The Boer position occupied a frontage of 10 miles, extended from Leewkop 2 miles north of the river to Seven Kopjes 8 miles to the south and overlooked the whole country. The attack to the north tended to move outwards in an attempt to outflank Leewkop. There was consequently little to pin the Boers to the ground before the outflanking attacks developed while the disaster at Paardeberg had shaken their morale. The action petered out in a field-day manoeuvre. As a move on the chess board, it succeeded admirably and opened the road for a further advance on Bloemfontein. It failed, however, to inflict casualties or weaken in any way the

Boer power of resistance. Good horses might have turned the Boer withdrawal into a rout but the British and Australian remounts were not yet acclimatised after their long sea voyage and the chance of decisive victory had gone. That part of the 6th MI was confined to galloping and holding a series of small *kopje*s on the extreme left flank.

After Poplar Grove, Martyr's Brigade took over the protection of the drifts at Ostfontein across the Modder River, there to await the arrival of a convoy from Kimberley. This convoy of 500 waggons arrived on 11 March bringing with it Lieutenant Clowes who had been left behind at Maple Leaf camp with surplus kit and who was shortly to be posted to the Gordon Company. For the next six days the 6th MI continued to act as escort to the convoy; till they were able to hand it over to the 7th Division at Bloemfontein on 17 March, St Patrick's day. During this time, they had wearisome time and no fighting though the guns could be plainly heard ahead in the Dreifontein action on 12 March. The oxen of the company were worn out with the forced marching and progress was slow with four or five hours alternate trekking and halting, while at night when at rest outpost duty was very heavy.

Arriving at Bloemfontein, the 6th MI moved into camp at Brands Kop, 2 miles from the town, and there followed a welcomed ten-day rest to reorganise. Parades during this period were filled in by some much needed riding school and other mounted infantry instruction. On 26 March the Essex Section was transferred from 'D', the Gordon Company, to 'C', the Wiltshire Company, with whom they remained until the end of the war.

On 28 March, as a temporary expedient, the 6th MI were brigaded with the New South Wales Mounted Rifles and the 8th MI under Lieutenant Colonel Le Gallais to take part in the forthcoming action at Karee Siding, 20 miles north of Bloemfontein. With the brigade was a section of pompoms, then a new weapon to be used in action by the British Army for the first time. These Vickers Maxim's firing a 1lb shell in rapid succession had been adopted by the Boers with other German armament. Their use speedily spread to the British Army. Very effective in scattering groups of horsemen, their effect was rather one of moral than material damage. Moving off on the morning of 28 March, Le Gallais' brigade marched through Bloemfontein in the northerly direction. Glen, 15 miles north of Bloemfontein, was a lovely valley studded with trees and numerous farmhouses through which meandered the Modder River. It was reached in the afternoon. The brigade was on the point of pressing on as firing was reported a few miles ahead when sudden orders came in to move into camp.

Next morning, they moved off at 5 am to take part in the Battle of Karee Siding. General Tucker was in command with French and the cavalry on the left, Tucker's 7th Division in the centre and Le Gallais' MI brigade on the right. The Boers put up a stubborn resistance, but the hardest fighting fell to the infantry in the centre, where the Boers held a formidable position athwart the Karee Hills. Moving on a wide front over the open *veld*, Le Gallais pressed on towards a low ridge undercover of his pompoms. This proved to be the lightly held outpost line of the Boer left flank. The presence of pompoms puzzled the Boers at first, so came down to meet them under the impression they were their own men. The Boer scouts scattered as the fire increased instead of stopping as they had expected. The ridge was speedily occupied by the MI who now came under the shellfire of the enemy guns. On occupying the ridge, the MI were met by a crowd of native men and women gesticulating and waving handkerchiefs. These proved to be the inhabitants at neighbouring *kraals* who had been collected by the Boers to prevent them from giving information to the British. An hour later the advance was resumed to another ridge in conformity with the infantry advance. At the foot of the ridge stood a farmhouse flying the white flag and in the doorway a Dutchman calmly smoking his pipe. He proudly produced a pass from Lord Roberts to show he had not been engaged in the enemy fighting forces!

After another long pause two other ridges were galloped and captured in succession. At this point a Cape cart with two horses were seen on the *veld* in the rear, apparently, by its erratic course, making some attempt to dodge bullets. On its arrival out jumped the well-known figures of Rudyard Kipling and Bennett Burleigh thirsting for news as only war correspondents can. Kipling produced a bottle of whisky and asked a sergeant and private to have a drink. He was somewhat agreeably surprised when both kindly, but firmly, refused. With this little rebuff, let us trust the celebrated author felt that all tommies were not the same boisterous, hard-drinking characters he had graphically depicted in his *Soldiers Three.*

A couple of Boer guns were now bought to bear on the captured ridge and the infantry dismounted, falling back with their horses under cover of the brow of the hill. As the fire slackened the men again mounted and, led by their officers, swept over the brow of the hill, down the opposite slope and onto the last ridge held by the enemy. They were just in time to see the two guns being withdrawn at a gallop in the direction of Brandfort. The battle, which had lasted all day, was now over and the Boers beat a hasty retreat in the failing light. The British casualties, all told, amounted to over 200 and the Boers had also suffered heavily but, once again, the cavalry on the left, hampered by difficult ground, failed to cut off the Boer retreat and so obtain a decisive victory.

The Boer forces took up fresh position in front of Brandfort seven miles to the north. The outposts of the two forces occupied two ridges overlooking a fine valley some 3 miles across with numerous farms. These the Boer and British patrols visited alternately since each visit could be clearly seen by the opposing outpost line. Only when the British attempted any outflanking or moved to beyond the neutral zone did the Boers retaliate with shell or musketry fire. Outpost duty was very heavy but, otherwise, the next ten days passed quietly enough with an occasional light skirmish when out foraging. On 8 April, Le Gallais' brigade was relieved by the 2nd Mounted Infantry and CIV and moved back to Bloemfontein where the mounted infantry were being reorganised to take part in Ian Hamilton's march as right flank to the general advance.

The action at Karee Siding marked a definite step forward in mounted infantry warfare. Hastily organised, thrown together without previous experience or training, badly equipped and at the outset handicapped by a great shortage of officers to act as instructors, the advance along the Modder River to Bloemfontein was little more than a rough preliminary schooling in the future duties. From now onwards the mounted infantry was to take its place as a special branch of the mounted forces. In the reorganisation at Bloemfontein there was a general smartening up of equipment and uniform and it is from this period that the special MI badge for the helmet was instituted. Apropos of the smarting up, a good story is told of one of the company who strolled into a barber's shop one day. He asked that his locks (which were by now more appropriate to cavalier of King Charles' day than the nineteenth-century soldier) might be removed. With a gentle tap on the shoulder, he was gently reminded by the proprietor that the saloon was for officers only. 'I am really very sorry,' murmured the man apologetically, 'but for heaven sake sell me some curling tongs!' The month spent by the army in Bloemfontein before the general advance could take place bore little resemblance to war. An officer of the Bedfords, newly arrived in the country with a draft and now posted to 'A' Company of the 6th Mounted Infantry, thus described his impressions on arrival.

'At Bloemfontein was established Army headquarters. All the public buildings had been taken over by the British forces and the Union Jack was floating from the flagstaff. The club was full of officers in every variety of uniform and from every part of the Empire. In the evening the bands, drums or pipers played in the Market Square, where Lord Roberts and his staff could be daily seen talking to groups of officers. The inhabitants, who were mostly women, being the wives and daughters of the Boers on commando, seemed to take the situation quietly and were on quite good terms with the British and colonial soldiers. Except for the occasional sound of distant gun and rifle fire from the direction of the advance outpost line to the north, there was little indication of war. Disease unfortunately was rife and there were many cases of enteric fever in the hospitals.'

On arrival back in Bloemfontein the 6th MI went into camp at Rustfontein about a mile out of the town. On 11 April they were inspected by General Sir Edward Hutton, at the time the Inspector General for the MI. General Hutton eulogised the good work done by the 6th MI and, remarking on their ragged condition, hoped to see them in a day or two more respectable in turnout.

In the new organisation Lieutenant Colonel De Lisle was given command of the II Mounted Infantry Corps with Captain Robert Fanshawe[13] of the Oxford Light Infantry as his staff officer. The II Corps consisted of the 6th MI Battalion, the New South Wales Mounted Rifles, a pompom section and supply and hospital detachments. The II Corps, which finally became an independent column, continued substantially the same for the next year or so. The II Corps, MI were attached to General Ridley's brigade.

It should be explained that at this period the mounted infantry and colonial mounted regiments were brigaded into corps to distinguish them from the

[13] **Major-General Sir Robert Fanshawe, KCB, DSO** (5 November 1863 – 24 August 1946). At Godley's invitation he attended the 1905 MI Concentration and Cavalry Manoeuvres held at Churn; see chapter 10. In the First World War he commanded the 48th (South Midland) Division from 1915 -18.

regular brigades of cavalry and infantry. They were, however, from time to time, referred to as brigades, thus causing some confusion. Later on, the term column was used irrespective of the actual composition of the different forces engaged. The 6th MI was now brought more or less, up to strength and additional officers posted to make good the casualties so far suffered. The following is the composition of the regiment prior to the departure of Ian Hamilton's force:

Commanding - Captain Pennefather, Welsh Regiment.
Adjutant - Lieutenant Anley, Essex Regiment.
Quartermaster - Captain Shaw, Wiltshire Regiment.
'A' Company Bedfords - Captain Nicholl, Lieutenant C.P. Strong.
'B' Company Welsh Regiment - Lieutenants Derry and Ferrar.
'C' Company Wiltshire Regiment - Captain Tufnell, Lieutenant R. Cockburn.
'D' Company Gordons - Captain R. Gordon, Lieutenant Clowes
Attached - Captain Honourable Gerald Cadogan, Lieutenant Oppenheimer (West India Regiment).
Medical officer - Doctor R. Kaye.

Of these fourteen officers only four were to see the campaign through with regiment.

As so often in war, final orders for the advance came in a hurry. On 20 April the 6th MI took over the outpost line three miles to the southeast. Spare stores and kit were left stacked in the standing camp at Rietfontein and the regiment moved on very light scale with the baggage to follow. On the evening of 22 April the regiment received orders to join II Mounted Infantry Corps headquarters at Springfield, northeast of Bloemfontein and to start at once. Hastily saddling up with what light baggage they had with them, they moved off in the twilight led by a local guide from Rimmington's scouts. After four hours marching in the dark, Springfield camp was at last reached. The general advance was to commence the next day but kits were still stacked at Rietfontein! Most of the heavy baggage was never seen again. Meanwhile Captain Gordon with a section of 'D' Company and a section of West Australian Mounted Infantry had been detailed on 20 April to commandeer horses in the neighbourhood of the Modder River, west of Bloemfontein. On returning successful from the foray four days later they too discovered that II Mounted Infantry Corps headquarters had moved to Springfield and were then starting on the general advance. The next day Captain Gordon, with all available details of the 6th MI and Australians belonging to the II Mounted Infantry Corps, were attached to Colonel Mitford's Colonial Horse. They were hurriedly despatched to Boesmans Kop, halfway to *Thaba Nchu*, on convoy escort duty and only the lighter kit from the stacked baggage could be taken with them. The remainder shared the usual fate of unprotected stores on service. If kit was to be light, however, there were to be many other compensations as compared with their comrades in the infantry working alongside them.

Moving off shortly after daybreak, and with a two-hour halt at Bushmans Kop, which was found unoccupied, the drift over the Modder was secured without much serious fighting by dusk. All around at Sanna's Post was the evidence of Broadwood's disaster a fortnight earlier - charred ox waggons, exploded shrapnel, shells, burnt limbers and ammunition boxes, as well as carcasses of horses and oxen, a testimony to the severity of the fighting.

From the outset of the march the MI were to earn a reputation as efficient moss-troopers[14]. Before going into camp from the drift, a farmhouse was found well-furnished and which residents had evidently left in a hurry. Butter, cream, hens, a pig and forage were all there for the taking. The 9lb or 10lb of butter was rolled into a pillowcase and stuffed into a nosebag; the pig was slain and strapped across the wallets of one of the men. Chickens were similarly strapped across the other troopers' wallets and the oat straw loaded onto a Scottish cart. The company returned to bivouac. 'Just like a walking provisions store,' as a facetious colonial trooper remarked as he passed by.

On 25 April the 6th MI was left flank guard to the force entering *Thaba Nchu* and had quite a sharp little engagement where the road leads over a

---

[14] Moss-trooper is one of a class of marauders who infested the 'mosses' of the Scottish Border in the middle of the seventeenth century; a border freebooter. The phrase "moss-trooping" is recorded in Scott's "The Lay of the Minstrel": "A stark moss-trooping Scot was he …" Deneys Reitz, author of several books including "Commando", uses the term in describing the activities of the commando as they go from one village to another.

depression between two *kopje*s, known as Israels Poort. The place was strongly held, and the advanced guard of Marshall's horse came under a heavy fire. Marshall, one of his officers and several of his men were killed, while a number of officers and men were wounded. After some brilliant skirmishing by the MI, the Boers' right flank was turned and their rear threatened, after which they retired on *Thaba Nchu.* It was now getting late and as *Thaba Nchu* was 8 miles further on, the force went into bivouac for the night. *Thaba Nchu* was entered next day after little desultory firing, and Captain Gordon rejoined with Mitford's convoy.

Among the hills to the south lay the main body of the Free State army, it having discreetly retired northwards from Wepner and Dewets Dorp on scenting the threatened invasion.

On the afternoon of 26 April the 6th MI were directed to take a forced march over the southern slopes of Thabanchu Mountain to cut off a Boer convoy trekking north. A two-hour ride over fearfully rough ground ensued but the convoy was already out of sight. The MI returned to bivouac long after dark and 14 hours in the saddle. Sir Leslie Rundle's column and one of French's cavalry brigades now came in from the south, and on 27 April the 6th MI moved back to Israels Poort to protect the rear and L of C with Bloemfontein. Here they found Meyrick and his company and the two Gordon companies were able to foregather and compare notes.

A good deal of sharp skirmishing took place during the next three days to cover the withdrawal of the Boer forces to the north through the broken country of the Ladybrand District. On 30 April Ian Hamilton's troops moved north to force the Hout Nek Pass as the first move in the general advance. The Boer position was a strong one. The left rested on the high hills which form the outer spurs of the Thabanchu Mountains, among which a long-range Creusot gun had been placed. The centre was filled by several low ridges through which ran the gap flanked by acacia scrub, known as *houtnek*, while on the right loomed the high precipitous, flat-topped *kopje* known as *thaba* (Basuto for a table mountain). Guarding the gap between the hills like a sinister sentinel stood the thin column of basalt rock called Sout Pillar after Lot's wife.

De Lisle's II MI Corps formed the right flank while the cavalry endeavoured to force the actual pass. A withering fire from the low ridges soon drove the cavalry back with some loss, while the right flank also got into touch with the Boer. Picquets and a general engagement commenced. The Creusot gun opened fire on the 1st line transport and ammunition waggons of the I Mounted Infantry Corps forcing them to retire out of range. The 6th MI now deployed, the two leading companies 'A' and 'D' moving at a gallop to establish a footing on the nearest ridge. 'A' Company on the right made for a group of farm buildings and native *kraals*, under cover of which they dismounted under shell fire and then skirmished up to where an outcrop of rock at the top of a low ridge gave reasonable cover and a field of fire. Several of their horses were hit by pompom fire while making their way to cover. One man being knocked down, though untouched by the force of a pompom shell which killed one of the horses he was leading. 'D' Company was less successful. They and a body of enemy horsemen made a simultaneous gallop for the ridge immediately in front of them. It proved, however, to be already held by Boer skirmishers who opened up at 50yds range, fortunately too excited to bring their rifles to their shoulders. The company had to retire at the gallop under a hot cross rifle fire from the heights on either side as well as coming under shell and pompom fire. Cover was reached 1000yds behind a low dam and some cattle *kraals*. The company lost several men and horses in the withdrawal and was very lucky in not suffering heavier casualties.

It was now the turn of the infantry to come into action. Gordon Highlanders on the right, Canadians in centre, Kitchener's Horse on the left could be seen advancing against the left of Thaba Hill. As they came into view the Boer guns, Krupp, captured RHA 12 pounders and Creusot, opened fire from Hout Nek ridge to which the British gunners made a vigorous reply. The rattle of musketry and the louder boom from the guns was continuous while shrapnel could be seen bursting over the scrub-covered top of Thaba Hill. Away on the right flank the whole progress of events could be clearly seen as the Highlanders and the Canadians pushed and struggled their way up and through the undergrowth of the steep face of the *kopje* and established themselves on its outer lip. To the onlookers, as darkness came on, the effect of the bursting shells and the spits of rifle fire of attackers and defenders was weird in the extreme. As soon as it was dark the Shropshire Light Infantry and East Yorkshire Regiment came up and relieved the mounted men all along the line. A cold and uncomfortable night was passed by the waggon, as no fires were allowed and only biscuit and cold water was available for supper.

Before daybreak the mounted troops were back in their former position and as soon as it was light the battle recommenced. Shortly after daybreak the distant sound of cheering could be heard from Thaba showing the Highlanders were on the move. By 8.00 am Boers could be seen through glasses, running back closely pursued by kilted men with fixed bayonets who were dodging about among the bushes while the shells of the Boer artillery could be seen heavily bombarding the reinforcing infantry ascending to the aid of their comrades on the plateau. Broadwood's cavalry were, meanwhile, working round the left flank, while the two 6in Cow[15] guns firing over the advancing infantry were putting the finishing touches to the assault. By 11.00 am the whole plateau was clear of Boers and the enemy were in full retreat.

At midday the 6th MI were withdrawn to their waggons for a welcome meal for the men and horses and after a very brief rest they were again mounted and sent to join in the pursuit. Only a nebulous rearguard was encountered, and these were hastened on their way by the battery, to which the MI were escort. Bivouac was pitched two miles north of the position where Bruce Hamilton's 21st Brigade, who had come out to reinforce, were already encamped. 'A' Company, meanwhile, were left in charge of a Boer hospital at a farm, most of the patients being foreigners. The doctor in charge was a Russian with an English-speaking Dutchman as an assistant. Among the wounded were two Irish Americans, two German artillerymen, an American, two Swedes and two Transvaal burghers all more or less badly wounded. Also among the Red Cross assistants were two English-born but naturalised Transvaalers. The two latter were offered places in the hospital but declined, urging they would be more use to wounded prisoners in the Boer lines. Another Londoner, a Free Stater of 27 years standing in the country, with a farm near Kroonstadt, was later found exchanging the role of Boer Red Cross assistant with that of dairyman to British troops at Kroonstadt and finally becoming driver to the British Army. At sunrise on 3 May the move forward was resumed. For the first four miles by open country, after which some stony ridges loomed in view, and with them shooting commenced. The 6th MI galloped a couple of these ridges in succession on the right flank under a heavy but erratic fire and cleared the advance for another few miles. Camp was pitched at Osobolafontein after

[15] Cow guns – artillery hauled by bullocks.

a 14-mile march. Cavalry and MI with Smith Dorian's and Bruce Hamilton's brigades all being in close proximity.

The next day, 4 May, fighting started almost immediately after leaving camp. The Boers were holding a strong position athwart the Tabaksberg consisting of a halfmoon ridge strongly protected on their left by a steep wooded *kopje*. The 6th MI were on the British left with the colonials in the centre and cavalry on the right. Immediately after getting into touch the Boers opened fire on the centre and right and 'A' Company was ordered to feel for their right flank. Taking advantage of the cover of a farm and some native *kraals* they cautiously moved to a low wooded hill on that flank. Hitherto the Boers had reserved their fire but on an attempt being made to outflank the hill the Boers suddenly opened fire at close range forcing the leading section to retire at a gallop. The company was now somewhat isolated and 'D' Company was sent up to reinforce, at the gallop, under a shower of Boer bullets, and Colour Sergeant Grassick was hit as he dismounted on the position. With other MI companies coming into action, 'A' Company was ordered to rally on the New South Wales Mounted Rifles who were holding a ridge to the left rear. The withdrawal from the firing line was successfully carried out by galloping back in sections, the men at 40-yard intervals.

De Lisle now brought up the whole of II Mounted Infantry Corps to reinforce the centre of the attack. Further to the rear the infantry had started to deploy. The cavalry, supported by the 5th MI working with great dash, had now established themselves on the outer spurs of the position on the right. The Boers started a counterattack but at this moment Captain Rose led the Household Composite Cavalry Regiment in a charge over a low nek on the right centre of the position. Captain Rose was mortally wounded and twelve or fourteen of his men hit but they got through followed by a Royal Household Artillery Battery and the rest of the cavalry. The site of the long, unbroken lines of advancing infantry completed the Boer discomfiture and they hurriedly withdrew guns in a north-easterly direction. De Lisle's corps joined in the pursuit, which lasted for another six miles. The Boers, however, had got a good start and De Lisle's corps encountered few of the enemy but much evidence of his hurried departure. Cape carts, blankets, rifles and ownerless horses were on the far side of the hill; the MI quickly annexed all serviceable articles and the men changed their saddles onto the fresher Boer horses as quickly as possible.

Bivouac was pitched at Welkom Farm on the little Vet River, five miles short of Winburg, and on the way, one of the officers secured a Springbok to supplement their rations. So ended a successful day which, at first, had promised to be another stiff fight like Hout Nek.

Next day the mounted troops moved round Winburg to the northwest encountering a desultory resistance en route and returning to Winburg, which the infantry had entered unopposed, in the evening. Firewood was always a problem on the march and the grandstand on Winburg race course on this occasion proved a veritable godsend to the troops. Even the winning post was uprooted while the judges' box was carried off in triumph by four stalwart Highlanders.

There followed two days' halt at Boomplatz, eight miles from the town, to which several bucks were secured by the various foraging parties, and then a short march to the vicinity of the Zand River, which was believed to be strongly held. The vicinity of the river was gained about midday and a fusillade greeted the advanced guard from several *kopje*s firing from its southern bank. The 6th MI were on the right flank with 'D' Company leading. At this point the river takes a bend to the south and 'D' Company galloped a ridge covering a drift across the river and faced by another ridge on the opposite bank, which was held by a couple of hundred Boers. Meanwhile, the Boer guns had opened fire from the high ridges on the north side of the river and an exchange of shell and rifle fire became general. Gradually, however, with no further British advance, the fire slackened and died down. In front of the ridge held by 'D' Company lay a tempting farm but all attempts of patrols to reach it were promptly checked by the Boers on the far side of the drift, who held the higher ground. Later in the day more than one inquisitive cavalryman wandering along in search of forage tried a visit despite warnings, only to get a nasty fright on nearing the farm, which in any case the MI had marked down as their special prey as soon as the opportunity should offer. Towards nightfall Ridley's mounted infantry brigade, of which De Lisle's corps formed part, was relieved by the infantry of the 21st Brigade on all but the extreme right flank. 'D' Company were, however, left on outpost on their ridge facing the upper drift.

On 10 May, the morning of the battle, the troops were up three hours before daybreak and 'D' Company was relieved by Kitchener's Horse, to join their MI Brigade. Concentrating with Broadwood's cavalry behind the infantry to await the course of events. The Derbyshire Regiment and the Cameron Highlanders of the 21st Brigade pushed across the Zand River before day broke, wading waist deep in the icy cold water, and surprised the Boer picquet asleep at the drift. Using only their bayonets they passed onto the *laager.* As soon as the occupants of the *laager* woke to the alarm, they retired immediately to broken ground in the rear, where they kept up a galling fire for some time. The British artillery had, however, been trained on the Boer rear positions the previous night and was only waiting for the signal of rifle fire to support the attack. The Boer main position was shortly afterwards carried by the Royal Sussex Regiment, at the point of the bayonet. The Boers made a desperate attempt to retrieve the position by a spirited counterattack on Kitchener's Horse at the upper drift and so turn the British right. Kitchener's Horse were hard put to it holding this attack and suffered heavily in casualties. Timely reinforcements, however, saved the situation and the Boers were driven back to cover across the river and on to their rear position. As usual the Boers did not stay to contest a position already half in enemy hands and withdrew with the easy mobility of mounted irregulars.

It was now nearly midday and the cavalry, supported by the II MI Corps, crossed the upper drift in an endeavour to cut off the Boer retreat. The cavalry had crossed the first rise in the ground in front of them and were sweeping round towards the enemy, when the latter opened fire on them with two of the captured 12 pounders RHA guns. The ground was favourable to a charge and a rush through would have recaptured the two lost British guns. Instead of pushing on, the leading cavalry squadrons wheeled round and came charging back on the MI throwing them into temporary confusions as they passed by.

A section of the second corps RHA battery, however, immediately unlimbered in the open and returned the fire. This miniature artillery duel continued for some moments in full view of the MI. The British gunners quickly found the range and silenced the Boer guns in splendid fashion. The MI formation had been somewhat broken up by the hurried retirement to the cavalry and on being reformed they were sent to the shelter of a *kopje* held by the Cameron Highlanders, who had been amused spectators of the whole performance. There was still a fair chance of capturing the guns by a vigorous pursuit, instead of which, the MI, for some unexplained reason, remained dismounted for nearly two hours before they were again allowed to advance.

Then came news that a Boer convoy had been sighted. No time was lost in mounting and away they went after the cavalry with the Boer convoy plainly visible some 6 miles away. First a trot, then a canter and then a gallop, cavalry, MI and a battery of artillery tore over the intervening ground. Then a Boer gun opened fire and rifle shots were exchanged as soon as the British came within range. The cavalry halted, apparently uncertain what to do next. De Lisle, however, leading his MI corp went right through the cavalry and soon left them behind. The fire from the convoy getting hotter, the MI dismounted and closed in on foot. The Boer gun having done its best hurriedly limbered up and was tearing over the ground not half a mile away when it suddenly disappeared among the slow-moving ox waggons and was seen no more. When the tail end of the convoy was reached the cavalry and guns were still nearly two miles behind. Darkness descending, the flying column bivouacked where it stood along with its captures, tired out after his eighteen hours of work. On 11 May the cavalry and MI moved to Ventersburg collecting many rifles and taking about 100 prisoners. Eventually they bivouacked at Blue Gum Spruit some 16 miles from Kroonstadt, rejoining the main body of Hamilton's force at sunset.

Kroonstadt, 20 miles further on, was entered on 12 May in the wake of other troops and bivouac pitched at Kroomspruit 5 miles south of the town. Here Ian Hamilton's whole force was gathered. Maxwell's and Smith Dorian's infantry brigades; II, V, VI and VIII Mounted Infantry Corps; Broadwood's cavalry brigade; divisional artillery and the two 5in guns. At Kroonstadt a two-day halt was made, including church parade and a visit from Lord Roberts who rode round personally to the various corps and brigades of his army. From Kroonstadt, Hamilton's force moved eastward to Lindley, 50 miles from the railway before again moving north.

Lindley was reached on 18 May, practically without opposition. The cavalry under Broadwood had entered the town the evening before, nearly surprising a Boer commando which had halted in the town and who had lost most of their baggage and supplies in getting away. Only one day's halt was made for supplies, and the march northwards was resumed early on 20 May. Trouble began almost immediately after leaving town. Not only were the advanced and flank guards constantly engaged in a running fight with Boer skirmishers, but the rearguard was heavily attacked. 'B' Company of the 6th MI were in action the whole morning on the right flank, and the cavalry men were held up for an hour before they could clear the front with the assistance of the Cow guns. All day long the sound of heavy fighting could be heard behind on the left flank. In the rear, half an MI Company of the Royal Irish Regiment, left to hold a *kopje* on rearguard, had been unable to extract themselves. They lost half their number in killed and wounded, and all their officers were wounded. Their horses had stampeded, and they were forced to surrender. Next day the whole II Mounted Infantry Corps were on rearguard with Colonel De Lisle in command. The rearguard was closely followed up the whole day, and the RHA Battery and two pompoms constantly in action. Casualties were, however, slight, one man killed, and one man wounded, both of them West Australian Mounted Rifles. Bivouac was only reached at dusk.

Heilbron was entered with slight opposition on 22 May, the 6th MI having, meanwhile, made a long detour southward pursuing a party of Boers towards Reitz. The rearguard was again followed up with only a few men lost because of wounds. From Heilbron, Broadwood, with the II Mounted Infantry Corps attached, moved northwest and, the second day out, camped on the railway three miles north of Lord Roberts' main column. Lieutenant Clowes and half of 'D' Company as escort took in seventy Boer prisoners to the main column. Shortly after daybreak on the morning of 24 May Broadwood crossed the line of march of Lord Roberts' force and off saddling three miles to the west, watched the column file by. After some hours halt, saddled up and set off on a 16-mile ride, an alternate trot and gallop till they reached the Vaal River at Boschbank Drift, which they made good. The 26th was spent in outpost covering the crossing of Ian Hamilton's force, which had now become the left column of the advance. The Boers kept the outpost busy, and an Australian outpost was captured.

On 27 May Broadwood's force moved to within sight of the Gatsrand Hills covering Johannesburg, meeting very little opposition on the road in spite of the ruggedness of country. In the evening General French and his calvary division came in and joined the column from the Western Free State. Early next day, on climbing the Gatsrand Hills, the chimney stacks of the Witwatersrand and Johannesburg burst into view. Below lay the Klip River valley and, on the far side, the steep ridge of Doornkop where Doctor Jameson had surrendered. The day was spent in feeling for the Boer position.

Away to the left the reconnoitring patrols of French's cavalry and Hutton's MI found themselves held up on every gap or ridge as they tried to feel their way through. Boer and British guns were in action all day and, by nightfall, French, who had established himself at the north end of the Klip River marsh, was obliged to withdraw all his mounted men under cover of darkness over the bridge and causeway, which were the only means of crossing the marsh. Next morning, on 29 May, on again advancing the MI found the Boer artillery trained on every approach, especially the bridge, and any further advance was held up for the infantry attack. As the infantry deployed the mounted troops were withdrawn and French reinforced by Broadwood and De Lisle started on a flanking movement towards Roodspoort on the Potchefstroom Railway. Bivouac was at Doornkop alongside the victorious infantry; the 6th MI fared well that night for 'A' Company on an outer flank had struck a farm with an old Boer Commandant related to the Botha family, who was on the point of departing with several Cape carts loaded with provisions. The carts were forthwith annexed and were further loaded up with fodder, potatoes, geese and pigs from the farm.

The next two days were spent at Florida, a wealthy suburb of Johannesburg, and an organised search made for provisions for the troops. On the way in a patrol the 6th MI aided by some Australians stopped and captured a train starting for Krugersdorp and made prisoners of the passengers. The Sappers then blew up the rails to stop further outgoing traffic. While searching Langlag, another suburb, the MI came upon a miniature battle between mine employees from the native location and the white citizens of the place. Rifles and other weapons were being freely used. A bloodless charge of the MI using their rifle butts indiscriminately on the heads of either combatants speedily put an end to the fight. Rimmington, with his Guides, came up shortly after and complete order was soon restored. Four natives caught red handed with loot received twenty-five lashes each with a *sjambok*, which acted as a further deterrent to looting.

On 1 June the II MI Corps marched through Johannesburg itself and camped in picturesque suburbs north of the town, while a halt next day gave opportunity for much needed weekend shopping. On 3 June the march forward to the Transvaal capital was resumed, Lord Roberts in the centre, with Ian Hamilton and French on his left and other mounted troops on his right flank. Bivouac that night was on the Yokeskei River (*Jukskei* River) only 11 miles from Pretoria. Shortly after daybreak next morning on 4 June, Broadwood's column was underway again in a northwest direction. They had gone little more than 3 miles when they were recalled to follow the main road, and the rumour rapidly spread that Pretoria had surrendered. The rumour was further borne out by the fact that there had been no sounds of firing.

About 9.00 am all dreams of a triumphant and unopposed entry into the capital were rudely dispelled by the sound of heavy and continuous gunfire far away to the right. The next 7 miles were rapidly covered as Six Mile Spruit and several ridges were crossed where British artillery were already positioned, vigorously shelling the silent Boer forts. The continuous rattle of musketry grew ever more distinct showing that the infantry were already hotly engaged and, on topping a rise, the whole action in progress lay before them. Floating above a flat top *kopje* was the War Balloon, and a battery of howitzers in front were hurling the shells over a ridge behind which Pretoria presumably lay, while further to the left and nearer two Horse Artillery batteries were in action, bursting their shrapnel over a bush-covered slope which lay in front. Below, the long lines of advancing infantry covered the plain like flies. After a brief look, orders came to seize a long low ridge covered with scrub immediately to the front. Galloping their horses to the foot of this ridge without coming under fire, the MI dismounted and led their horses up the steep slope. After a stiff climb to the top there proved to be a second higher range 500yds further on. Immediately the men showed above the crest they became a target of a long line of Boer riflemen lying hidden among the bushes and boulders of the ridge. A pompom was accordingly manhandled to the top and the ridge plastered with shell. A long wait followed whilst the Boer fire gradually slackened, till Ian Hamilton, who with his infantry had gained an advanced position on the right, sent word to De Lisle that he could see the Boers retiring on the II Mounted Infantry Corps' front. De Lisle at once ordered the corps back to their horses. Leading them along through rock and scrub they worked their way, under cover of the crest of the ridge, in single file for some considerable distance to the left. Here they were able to descend by several precipitous paths through a cleft on the northern face of the ridge to the plain below. There were many casualties in horses owing to the bad going before the plain was reached.

Meanwhile, the Duke of Cornwall's Light Infantry had reached the ridge the MI had just vacated and were in a position to cover the mounted infantry as they mounted their horses and rode out of the scrub into the plain. The few remaining Boers on the ridge facing the infantry hurriedly took to their horses and made off. 'D' Company of the 6th MI were the first to get mounted and, led by Captain Gordon, dashed off into the plain at a gallop, the remainder of the MI mounting and deploying in one long line across the plain as fast as they could get mounted. After half a mile or more Captain Gordon could be seen to wheel to the right after some Boer waggons which were trekking as fast as they could to the town only two miles distant. The two last waggons were overtaken in the municipal boundary pillars; the Boer escort abandoned the waggons and teams and galloped off into town as the MI approached. The Gordon Company of the 6th MI were thus the first British troops to enter the outskirts of the town.

The 6th MI found they had successfully turned the Boer right. British howitzer and naval guns were bursting their shells in the town and around Klapperpoort and Wonderboom Forts. Some of the shells fell perilously near the mounted infantry, who had not yet been located. Night was fast approaching, so De Lisle decided to hold two small flat-topped *kopje*s about 2000yds to the north of the outskirts of the town. Captain Gordon, at the boundary, was keen to enter but the message came to stay where he was till bivouac was formed and then return with his captured waggons. De Lisle then sent off his ADC Captain Watson of the New South Wales Mounted Infantry with a flag of truce to Pretoria to interview the authorities. After a long wait he returned accompanied by the Transvaal War Secretary and the Mayor of Pretoria. Colonel De Lisle personally escorted these two officials over the hills to Lord Roberts' headquarters. Lord Roberts, however, refused to see them as it was then 2.00 am. However, he sent message that he would accept the formal surrender of Pretoria on the outskirts of the town at 9.00 am.

As soon as it was dark the electric lights in the town began to come on while a large blaze to the left, under Meintzers Kop, lit up the Birdcage where the British prisoners lay. Two or three Boer prisoners offered to guide the corps to the Birdcage with a view to their immediate release, but the corps were too much 'in the air' to deem it wise to try a risky night adventure. They likewise impotently watched two trains leave the station and steam eastwards along the Delagoa railway.

The contents of the two waggons captured by 'D' Company proved most acceptable to the 6th MI, bivouacking without waggons just as they stood. In one waggon were 100lbs of bread in sacks, some bags of rusks, a 100cwt of brown sugar and some tins of coffee and tea. In the other was a .303 Vickers Maxim complete with tripod, 300 13-pounder Krupp shells and cartridges and several bales of *Staats Artillerie* uniforms and greatcoats. The latter was soon dispensed among men who had long worn out their clothing and were almost in rags. With sunrise the II Mounted Infantry Corps, from their bivouac near Pietersburg railway, watched the infantry filing through various gaps in the ridges into the plain below and forming up on the outskirts of the town. At 9.30 am they heard Lord Roberts had accepted the unconditional surrender of the town. By noon the first of a long line of cavalry, artillery with their Cow guns, howitzers, field guns and pompoms, infantry battalions, and regiments of MI bringing up the rear were threading through the streets of the capital.

The II MI Corps, to their disappointment, were not allowed a part in the formal entry they were so near to accomplishing the night before. Exactly a fortnight was to pass by, and with it much hard fighting, before they were to tread the streets of the capital whose city boundaries they had been the first British troops to cross. Instead they were directed to hold a line of picquets four miles east of the town below Klapperpoort fort. Fresh disappointments waited them next day. On being relieved of picquet duty they were allotted a pleasantly wooded camp and promised a week's halt. On arrival of their waggons, however, and before they could start unloading, fresh orders came to saddle up at once and march to Irene on the railway, 12 miles south of Pretoria.

Ridley's brigade, joined by Broadwood and the 21st Infantry Brigade, did not reach Irene till after dark on the evening of 6 June. The advanced guard was somewhat puzzled to find a high barbed wire fence on reaching bivouac, but it was quickly cut and the force passed through to camp on the other side.

Captain Nicols wrote in his South African Journal that: 'As soon as it was light, desultory firing broke out and the awakened camp hurriedly tumbled out of its blankets thinking Boers were upon them. A most extraordinary sight then met their eyes. Several soldiers were pursuing a mixed herd of zebras, sable antelope, hartebeests, wildebeests and brindled gnu, and other antelope which were cantering round and round the enclosure in which the force was bivouacking! The force had inadvertently pitched their bivouac in Pretoria Zoological Gardens or Park. Some officers soon stopped the firing but not before two of the zebras had been badly wounded and several larger antelope killed. Later in the day the unharmed animals got over their fright and wandered about amongst the men and horses of the bivouac, being made great pets of by the soldiers.'

The move to Irene had been a manoeuvre for position in the forthcoming battle of Diamond Hill. Early on 8 June they bid farewell to their stable companions of the Zoo and moved as advanced guard to the main force, to Zwaavels Poort some 16 miles east of Pretoria. Passing through the narrow gap of the Poort the force bivouacked on the edge of a large fertile valley dotted with farms. 'A' Company was vanguard and whilst picqueting the Poort struck their first orange grove at ex-President Joubert's fine farm and made full use of the opportunity in filling their nosebags with the ripe fruit. The next two days were spent in feeling for the enemy position while foraging parties were sent out to collect supplies. One of these from the 6th MI that came too close to the Boers' picquet line was nearly cut off by a strong party of Boers, sent down to try to cut them off. They had some difficulty in getting the loaded waggons away and lost two mules in crossing a drift, but luckily there were no casualties among the men.

At daybreak on 11 June the advance began with the cavalry leading, riding into the mist in an easterly direction. The low hills held by the picquets were soon cleared and the calvary began to spread out fanwise over the plain. In the distance, 2 or 3 miles off, loomed the Boer position with a formidable range of high *kopje* extending for nearly 15 miles across the front, held by 7,000 men with Louis Botha in command. Covered with numerous entrenchments, gun emplacements and powerful artillery the Boer position was indeed a strong one. No sooner had the leading cavalry regiment, the 12th Lancers, got well into the open when flashes appeared from numerous points on the mist covered hills and shells began to pitch among the horses and men. Opening their formation slightly the cavalry pushed solidly on, taking little heed of the shelling. Then the rattle of musketry broke out and the 6th MI were ordered to advance. The ground at the edge of the plateau was rather broken with rocks and 'A' Company, in the lead, dismounted until they reached better going and then, after hurriedly mounting, galloped for the cover of some stone *kraals*.

From here the horses were taken to the shelter of some farm buildings 200yds in the rear.

The cavalry, in the meantime, had been manoeuvring on the plain to the right rear with their right flank thrown still further back, as a puff of smoke on an outlying hill to the southeast had betrayed the position of another commando of Boers with a gun threatening the rear. The cavalry seemed undecided what to do and, after a short delay, the men dismounted, holding their horses. The shells still continued to pitch among them but as they were percussion fused and only exploded on striking the ground, they did little damage. Neither horses or men appeared to take much notice even when casualties began to occur. 'D' Company under Captain Gordon had now reinforced 'A' on the right joining up with the cavalry.

Bullets were now whistling by, but no sign of the enemy marksman could be seen. The ground in front appeared to be quite level to the foot of the hills. The 12th Lancers, immediately on the mounted infantry right, now mounted and started heading for a gap in the hills in front, in line of squadron formation. As soon as they were well started the Boer rifle fire, which up to that moment had been slight, suddenly redoubled and a scattered crowd of black-coated men on horses unexpectedly emerged from a hollow in the ground. The Boers cantered out to meet the Lancers. It was an extraordinary spectacle. The cavalry leaders and scouts were already among the Boers and the horses of the cavalry were falling in all directions. In a few seconds there was great confusion as the leading squadron came into a perfect hail of bullets at close range. Some of the Boers stayed in their saddles and emptied their magazines at point blank range while others dismounted and were firing from under their horse's necks. The cavalry were unable to drive the charge home or return the fire as the British cavalrymen had not been taught to use their carbines in such an emergency as this. The Lancers at last wheeled and returned to cover, the Boers keeping up the fire as they retired. As soon as he emptied his magazine, each Boer wheeled his pony and galloped back to the *donga* for cover, there to recharge his magazine. As soon as the Lancers had cleared their front the MI opened on the Boers as they galloped back to cover. The *veld* was strewn with wounded or dead horses and men and riderless horses were galloping about everywhere. The leading squadron of the 12th Lancers had managed to stick some of the Boers with their lances but nearly all their men were on the ground wounded or dismounted.

The 12th Lancers formed up and tried to charge again but the Boers, having refilled their magazines and emboldened by their first success, came out again and forced the cavalry to retire under a second hail of bullets. Two guns of the RHA Battery now came into action and shelled all the hollows in the ground with shrapnel. Scattered parties of Boers commenced to retire to the hills in their rear. Shell fire from both sides continued and away on the British left in the direction of Pretoria, an almost continuous cannonade could be heard. Infantry began to arrive on the scene and were reinforcing the line of the MI and cavalry all along the front. In the evening the British field guns had plastered the slopes with shrapnel and the Boers were forced further back. An attempt was made under Colonel De Lisle's orders to gallop forward and occupy a farm about 1000yds to the front. The moment the MI appeared on the open plain shells came shrieking from the hills and the leading company came under hot rifle fire and the line was forced back again. That evening the MI were back in their old lines by the *kraals*, very little, if any, progress having been made. No fires could be lit and the baggage was not up; however, they got a native woman in the *kraals* to cook some fowls and kids in a hut for them. The baggage guard itself had bumped into a Boer picquet on the right rear, which the baggage guard, aided by the Derbyshire Regiment on rearguard, had to attack and drive off their *kopje* after a sharp little engagement.

As far as Broadwood's mounted forces and the 21st Infantry Brigade were concerned, the day's work had proved disappointing. A few Boer picquets had been driven in but at a cost to the 12th Lancers of Lord Airlie, their colonel, Lieutenant Wright and seventeen other ranks killed, besides casualties in wounded and horses. The Boer main position on Diamond Hill was still intact. The fighting on the left continued well on into the night. As soon as the sun appeared over the eastern hills came the Boer guns saluting the daybreak. A patrol sent out at dawn reported the farm in front still held by the Boers. Meanwhile, a staff officer looking for De Lisle reported two 5in Cow guns were on their way up and that the infantry were about to attack. The artillery on the British left commenced to pour in a continuous shower of shrapnel on the lower slopes of the hills in front and presently they caught sight of the infantry in extended order advancing across the plain. The attack had begun in earnest. The cavalry were standing to their horses on the right and could see that the Tigerpoort ridges were held by their own men. The rattle of musketry rose and fell throughout the whole day without ceasing; every fresh advance was recognised by the sudden increased volume of fire. Men of the Derbyshire Regiment came up and passed through the lines. They were the right-hand regiment of Bruce Hamilton's brigade in the attack. Colonel De Lisle now ordered the mounted infantry to leave their led horses behind in charge of the 'Number threes' and advance on foot at a fifteen-pace extension, thus prolonging the lines of the infantry brigades to the right. The corps was directed to carry out a frontal attack on the hill from which the two Krupp guns had worried them so much the day before.

A pompom opened fire but there was no sign of the Krupp guns. When lying down after each advance the high grass of the *veld* gave fair cover from view. As they neared the foot of the hill, single figures could be seen running or riding up the hillsides and disappearing over the skyline. The musketry fire on the left now became a continuous roar and, as soon as the 5in Cow guns opened fire, a cheer went up from the whole length of the line. A sudden flash and a cloud of green smoke and dust marked where each lyddite shell had burst. The Boer guns replied at once but their shells fell well short of Tired Tim and Weary Willie, as the two Cow guns were affectionately called. The corp pompom under Captain Sperling was in action alongside the MI till a Boer gun got the range and forced him to retire. Another Boer gun was enfilading the infantry line and the din of battle was deafening. Weary Willie and his mate were pounding the top of Diamond Hill and the activity of the Boer on the ridge increased in restlessness. The two Krupp guns were now silent and there was no doubt the Boers were 'on the go'.

The attack was pushed on and the British could see that the CIVs had gained a footing on the slopes of the hill away to the left. Moving steadily forward, led by De Lisle, they at last reached a large farm surrounded by a garden and several enclosures planted with fruit trees. Skirmishing through there they found themselves close under Diamond Hill. Bullets were now coming thickly and four scouts were ordered forward to choose a way up the hillside. They soon signalled that all was clear from the rocks above them. The British then advanced in a general line, the supports closing up gradually. The reserves, consisting of the New South Wales Mounted Rifles, could be seen cantering behind over the plain they had just left. As the mounted men neared the foot of the hill the ground all around them was knocked up by volleys of bullets fired at them by the Boers on the crest above. The Australians broke into a gallop and closed in to the foot of the hill where they were more or less under cover. As soon as the reserves came up, they dismounted and reinforced the line and all commenced to scramble up the rocky slopes, each man picking his way and halting near the top of the ridge to allow those behind to close up. Scouts were sent ahead to the crest line itself and found, as so often in the previous fighting, that the crest was overlooked by another, 150yds in the rear, which was strongly held by the Boers. The line cautiously made their way to the crest line and both sides fired at any movement on the opposing crestline. Only the bare, flat hilltop lay between the two lines of rifleman but any attempt at a forward move was at once checked by rifle fire. A captain and a private of the New South Wales Mounted Rifles were both immediately shot dead on leaving cover. One of the Boers, who fired, was also shot dead as he slightly rose from cover to take aim. There was nothing for it but to wait till the evening's resistance slackened under shellfire before the order could be given to fix bayonets and charge. As the afternoon wore on the 21st Brigade, reinforced by three battalions of Guards, had also gained a footing on the main Boer position further to the right. The order to charge was not given till the sun slowly sank behind the burning hilltop, the grass all on fire from the flash of the bursting shells. Twenty minutes after nightfall, fixing bayonets the whole line rose and scoured the hill. The Boers had gone under cover of the smoke and flames of burning grass. Far away in the distance a train could be heard puffing away up the valley where the mountains of the Eastern Transvaal were to afford a last organised stand against the invading army. On gaining the position there were pools of blood, piles of empty Mauser cartridges and the usual detritus of a hard-fought fight but only three dead and two wounded Boers. The Boer, like the Indian tribesmen, carried off his dead wherever possible. One unwounded Boer was guarding his dead brother's body, risking death or banishment to secure a decent funeral. The body was buried with full military honours alongside the British fallen. Of the infantry, the Derbyshires had suffered heaviest but to them fell the proud distinction of capturing the only gun taken in battle. The II Mounted Infantry Corps casualties were relatively small, one officer and one man of the New South Wales Mounted Rifles killed and some twenty or thirty wounded all told, of which the 6th MI accounted for six.

Outpost were found by the New South Wales Mounted Rifles and the corps was withdrawn to the foot of the hill where their waggons and horses had already arrived. Fires were already lit and preparations for dinner in progress. The farm, to which reference has been made, providing goats, sheep and geese for the men and turkeys for the officers. Diamond Hill was the only action in which the mounted infantry fought throughout as infantry. The account of the events leading up to it has been taken from *A South African Journal* by Captain Nicol, who commanded 'A' The Bedford Company of the 6th MI at the time.

Early next morning 'A' and 'C' Companies of the 6th MI relieved the New South Wales Mounted Rifles on picquet and, at about 9.00 am, moved off as an advanced guard in the direction of Elands River railway station. The corps moved as right flank to the force and marched by a road which appeared to run parallel to the railway. The railway lay in a fold of the ground and somehow the leading guide missed it and lost touch with the main column. The leading MI was ordered to go on ahead to a *kopje* to see if anything of the main body could be seen and what sort of country lay beyond. The leading scouts ran into a Boer picquet and nearly captured it. A shot was fired and the Boers bolted down the hillside leaving their horses. The MI dismounted and Captain Gordon with 'D' Company came up to reinforce and riding through followed the Boers up. The *veld* was on fire in places and scattered Boer horsemen were seen cantering over the plain ahead. The Boers were firing the grass in all directions and shooting wildly from the saddles to delay the advance. Closing in upon them, as fast as their tired horses would allow, the MI, to their astonishment, came upon a huge Boer *laager* covering the ground for a considerable distance on both sides of the railway and moving aslant their line of advance. They had struck the left wing of the Boer army, partitioned the day before into two portions and now endeavouring to join up with its right wing. There was some broken ground between the MI and the convoy, of which the Boer rearguard took advantage to dismount and fire. A ridge northeast of Bronkhurst Spruit station was, however, gained from which the remainder of the Boer *laager* under Louis Botha suddenly sprang into view and only 1,000yds away. Flinging themselves from the saddle the MI started to blaze as hard as they could.

For some time panic reigned on terrified burghers who thought they had been overtaken by the British Army. Grazing horses were hastily seized and saddled, mules and oxen inspanned and started off while the Gordons, reinforced by the Australians, blazed away as hard as they could. Finding that the advance was checked, the Boers were rallied by their leaders and took up a position to cover the retirement by the waggons as each was rapidly inspanned. The Boer artillery of the rearguard also came into action against the ridge. Captain Nicol of the Bedfords, who had remained guarding the right flank, now saw a body of 200 Boers advancing round the right flank while another party was advancing against the Gordons and West Australians to the left. The whole party in action were only some 400 strong and Captain Nicol, handing over to Lieutenant Strong his subaltern, galloped back to De Lisle to explain the situation, which was growing critical. De Lisle, realising they were entirely separated from the main body and having no desire to defy the whole Boer Army alone, at once ordered a retirement. The Boers were none too certain of the situation so contented themselves with a parting fusillade and both sides rapidly drew apart, glad to be quit of each other's company. An easy 10 miles brought them to Elands River where they rejoined the main body in bivouac.

A day's rest followed and General Ridley, commanding MI Brigade, ordered a parade at which he said he had much pleasure in announcing that Ian Hamilton had made a special report on the way in which the 82nd Battery and De Lisle corps had carried out the attack on Diamond Hill and that Field Marshal Lord Roberts had also expressed his appreciation of the work of both units. A quiet two days' march to Pretoria followed and on 17 June the men were able to visit Pretoria for the first time. The visit was not one of long duration but the MI managed to make an impression all the same. It was Sunday and the shops in the town were closed but one of the largest stores, Beckett and Co, agreed to open for the incoming force. There were only two shop assistants who duly waited on the few officers first. Meanwhile, about a hundred hungry soldiers had forced their way in and the impatient men soon began to help themselves. It was a case of taking what you want and paying when you left. The scramble, not unnaturally, soon degenerated into a loot. While the highly respectable CIV contented themselves with biscuits, figs and other dainties the more boisterous cavalry and MI roamed the shop, invading other departments, exchanging old boots for new and otherwise annexing whatever took their fancy. Captain Fanshawe, staff officer to the II MI Corps, passing by and seeing a number of his corps in a queue along the street, went in and promptly had the doors closed. The 6th MI had made a speciality of tinned fish and with these showing prominently in their hands were allowed to depart without further questioning. Becketts Store, thirty years later, was still a flourishing concern in Pretoria, a prosperity perhaps partly due to it having been purveyor general to the Boer force.

Halting on the Monday for much needed remounts, on the morning of 19 June they were off early back to Irene. A few of their old friends at the zoo were still wandering about, but most had escaped to the open *veld* or had been killed by some party of hungry soldiers. Irene was to be the jumping-off place for another trek, this time to the Northern Free State to co-operate with Buller, who was now slowly working his way north from Natal.

Chapter 7

6th MI on the heels of de Wet and operations in Cape Colony

The new move had a double objective, namely to get in touch with Buller, who had crossed the Drakensberg and was moving northwest, and to cooperate with Roberts' army in the part of the ring of columns moving to surround Generals Prinsloo and de Wet in the North East Free State.

On 21 June Ian Hamilton's whole force occupied Springs, the centre of the Johannesburg coalfields. The march took place in drenching rain and the township was found deserted. Engine sheds were requisitioned and most of the troops bivouacked under a roof for the first time for many months. Office furniture, supplemented by lumps of coal, soon provided roaring campfires at which the troops could dry their sodden clothes.

Heidelberg was occupied next day with very little opposition but the picquets moving to the south of the town soon came into touch with the enemy and a sharp little skirmish lasting a couple of hours ensued with the Boers 200 strong on some *kopje*s to the southeast. Guns had to be brought into play before the enemy were driven off. During the fighting two of 'D' Company scouts had got too far ahead and mistook a party of khaki-clad Boers for Australians. Consequently they rode into a shower of bullets at point blank range: one horse was killed and its rider, Private Hutchinson, had his wrist shattered while firing dismounted. The other scout, with his horse badly hit, managed to get away. Four Boers left cover to secure the wounded man's rifle and saddlery and at first handled him rather roughly. Finding, however, that he was wounded they stayed on under fire bandaging his arm, an act the British could not see. The wounded man was then let go. This brave and chivalrous act of the Boers in bandaging him up, though under fire, undoubtedly saved Hutchinson's life. His arm was amputated that evening in hospital.

A halt of three days was made at Heidelberg during which Sir Archibald Hunter arrived with his division. Ian Hamilton had broken his collarbone in a riding accident and Hunter took over command of the entire force. On leaving Heidelberg the force split into two columns moving parallel to the Natal Pretoria railway. Buller's guns could be heard in action, though the road was clear on Hunter's front. The second day out, communication was established with Buller at Standerton and a force of 400 men sent to hold that town. Hunter's force then crossed the Vaal at Villiersdorp and right at Frankfort with practically no opposition. De Wet had only just left the town two hours before the advanced guard arrived. Some fifty prisoners from the Derbyshire Militia, which he thought best to release, were in the town to welcome the British arrival. A halt of two days was made at Frankfort to enable Hector MacDonald to join the force, bringing with him a big convoy of supplies from Heilbron. MacDonald's advanced guard of Lovat Scouts with due Scottish caution engaged Hunter's picquets before they would trust themselves to enter the town.

From Frankfort the force marched to Reitz, the march being delayed by *veld* fires and drifts. Reitz was entered on 7 July without opposition, but only a midday halt was made, and camp was pitched some seven miles further on towards Bethlehem. As a result of the many *veld* fires, waggons and their mule teams on two occasions were caught and burned, while in camp, horses were stampeded and Cape carts, tents and bivouacs were burned. Many of the *veld* fires were undoubtably caused by the Boers when the wind was favourable.

On 9 July, Hunter's force joined up with Paget at Bethlehem, which Clements and Rundle had entered two days before. Evidence of sharp fighting between Paget and the Boers had been evident from time to time in the marches of the past few days and strong Boer forces were still in the immediate neighbourhood.

The object of the three-week march south and ever-increasing concentration of columns now became apparent. At Mount-aux-Sources, the highest point in the range, the Drakensberg sends out a long arm 40 miles to the west, which forms the northern boundary of Basuto Land. Bethlehem, 30 miles northwest, guards the exit to this cul-de-sac, which is further shut in by a mass of high hills north of Harrismith, then in British hands and connected with the Natal railway.

A force of Boers, 6,000 strong under Prinsloo and de Wet, who had been operating on Buller's flank, had been gradually forced back by converging British columns till, too late, they realised that they had been trapped in this formidable corner of the great range which forms the backbone of South Africa. At the moment, however, the Boers were only beginning to realise their danger and there were constant sharp skirmishes between opposing reconnoitring patrols. In one of these, a party of Oxfordshire Yeomanry were ambushed from a hill known as Gibraltar, on the Senekal road, losing many casualties including an officer and ten men missing. Before morning on the next day, II Mounted Infantry Corps, reinforced by a battalion of infantry, moved out to the scene of the disaster. A strong cold wind was blowing in their faces across a burned *veld*. The column had barely crossed the first ridge beyond the outposts when, in front, a body of some 200 Boers could be seen galloping away. The isolated kop of Gibraltar was still three miles off. As the scouts reached the point it was found unoccupied but there was ample testimony to the gallant stand the Yeomanry had made. The missing men and officer were found dead; all had more than one wound and one man had fifteen empty cartridge cases beside him. On the kop itself were thousands of rounds of ammunition from Lee Metford, Mauser and Martini weapons, which the Boers had fired at the Yeomanry in the open *veld* below.

On returning to camp a long dusty cloud in the west showed the arrival at last of Hunter's long expected supply convoy. Next day, Sunday, 15 July, after church, the exodus of the troops from the town began. Continuing until nightfall, the various brigades set off to occupy several gateways or *neks* fringing the great basin of the upper reaches of the Caledon River into which the harassed Boers had now been driven. Paget's 20th Infantry Brigade reinforced by Broadwood's Cavalry Brigade was ordered to Gibraltar to try to block de Wet who was threatening to break out across the Senekal road. Ridley's MI brigade, meanwhile, remained in reserve at Bethlehem. At noon the next day whilst idly resting in bivouac, sudden orders came to saddle up and march in an hour's time. De Wet was on the move and during that night broke through the cavalry cordon with 2,000 men and three guns. Bivouac was pitched within sight of Paget's camp fires with every expectation of holding the enemy on the morrow. Moving off at sunrise, Paget's force was overtaken at 9.00 am, only to learn of de Wet's escape and to join with Broadwood in the pursuit.

Bivouac was pitched 10 miles south of Lindley, after a 25-mile march without coming into contact with the elusive burghers. For the moment anxiety was centred on a British convoy unsuspectingly on its way south from Lindley and early next morning a search was first made for it. The convoy was duly found hidden snugly in a hollow and had luckily escaped de Wet's notice. The convoy had nearly three months' mail for Ridley's brigade, which were hurriedly transferred and the pursuit again taken up.

Next day, 19 July, the rearguard of de Wet's force was overtaken in the afternoon, holding the Rhenoster on practically the same ground as the rearguard action which was fought by De Lisle's corps in the general advance to Pretoria just three months earlier. A sharp encounter followed known as Palmietfontein, which the Boer rearguard fought stoutly and were admirably served by their artillery. Several ridges in succession were rushed and captured, but still the Boer rearguard defied all efforts to break through, and when night fell de Wet's guns were still covering the withdrawal of his men, with the loss of only a few of his slowest waggons. Piet de Wet had shown himself as stubborn a rearguard commander as De Lisle. Casualties on both sides were heavy, some sixty all told on the British side, of which a third fell to the II MI Corps including the OC of the West Australians.

Handicapped by having to wait for the transport waggons, the march was not resumed next day till noon. In the afternoon the Boers laid a trap for 'D' Company on the right flank. Some waggons drew up at a farm and hoisted the white flag. Luckily the scouts discovered that though the farm was willing to surrender, the neighbouring *kopje* and *nullah* were strongly held by Boers. Shots were exchanged till dark when the company withdrew without attempting a capture. After dark the lights of the Boer *laager* could be seen from camp some miles to the north. Reveille next day was at 3.00 am but on reaching the railway near Roodeval it was found that de Wet had crossed during the night and had captured an ambulance train, which they denuded of comforts and then burnt. Traces of the mail captured by de Wet the month previously were evident everywhere in train wreckage and burnt letters and papers lying about the *veld*. By dark, touch with de Wet had been regained on the road to Vredefort.

Vredefort itself was found unoccupied but the Boers were on the hills nearby. De Wet had apparently again been cornered and some fifty casualties were suffered, driving in his outlying picquets. From 25 July, the operations were confined to those of an investment force till the infantry could be brought up and the attack pressed home. On 30 July came the welcome news of Prinsloo's surrender with 4,000 men. On 2 August, de Wet sent back his British prisoners, and on 4 August, Lord Kitchener arrived to take over command. All had seemed to point to de Wet and his 2,000 burghers joining Prinsloo in a few days' time.

Between besieged and besiegers in this part of the line there was a wide valley intersected by a stream, and small parties from the picquets indulged in snap shots at quail, duck, hare and buck under intermittent sniping by the Boers.

On 5 August Ridley's brigade moved to Parys, clearing out the Boer patrols on arrival. A quiet day followed but on 7 August heavy artillery fire, apparently at some force beyond the river, was heard from Metheun on the right. Gradually the truth leaked out that de Wet had once more got through. Too great caution had proved the British undoing. While Metheun, Broadwood and Ridley contented themselves with investing the Boers at a safe distance of 5,000yds, satisfied that no crossing could take place except at drifts, which were duly watched. De Wet had used the fortnight in constructing a new drift, naturally not under observation, covering this operation by feints of detached parties seeking an outlet to escape. As so often in the campaign, fear of casualties and their effects on the Home Front prevented a bold policy which would have obtained decisive results, so the war dragged on.

De Wet was not yet free, however, and close touch was kept with his rearguard. Lindique Drift was seized on the 9th after a race with a Boer regiment. As the dash was made for the drift, 200 of the enemy with two guns, detaching themselves from the main body, sought to prevent the crossing by occupying several bush-clad *kopje*s dominating the right bank of the ford. With a mad gallop down to the river, across it and then for half a mile up a stiff sloping valley, the corps reached the *kopje* first, breathless but triumphant. Bringing a rapid fire to bear on the enemy the latter turned and, retiring back for some distance, brought his guns into play. Eventually the pompoms under Captain Stirling won the day and the Boers disappeared over the hills. The British waggons crossed during the night. The Boer rearguard made some stand on the Gatsrand to enable de Wet to brush aside the weak forces holding the Klerksdorp line and blow it up before crossing. The Gatsrand was crossed early in the morning of the 12th and bivouac was pitched north of the railway but only for three hours when the march was continued by moonlight. Next day the brigade marched at 2.00 am and passed through Metheun's column, who had hitherto led the pursuit, while the latter were still asleep. Not long after, while it was still dark, a terrific outburst of musketry suddenly bought the column to the halt. The advanced guard had bumped into the enemy picquets. As soon as the firing began troops extended and waited for dawn. Daylight showed numerous ridges and *kopje*s thickly studded with bush and giving excellent cover to the Boer marksman. These were de Wet's rearguard holding the spurs of the Magaliesburg.

De Lisle was, unfortunately, hit in both arms in the early stages of the advance, though the casualties were otherwise small, and the hills soon cleared under shell fire. A little later the news leaked out that de Wet, on point of surrender, sent a patrol to Olifants Nek and found to his astonishment that Baden Powell was not there as expected. So, he quietly walked through, leaving his rearguard to hold the Nek till he was safely on the other side. The result was doubly disappointing for plans had been well laid. With Baden Powell on the north holding the passes through Magaliesburg, Smith Dorrien blocking any move to the east, Metheun to the west and Broadwood to the south, all the circle of investment had seemed complete enough for success to be certain. De Lisle, in spite of his wounds, continued to accompany his corps in a Cape cart instead of going to hospital and two months' later was able to resume command. With the failure of their coup the mounted troops expected to move into Pretoria as they had been more or less on the move for seven weeks on end. To their surprise, however, next morning, 15 August, found them suddenly ordered westwards as part of a combined force including Broadwood's cavalry, Ridley's MI brigade and Smith Dorrien's infantry brigade all under Lord Kitchener moving to the relief of Colonel Hore at Elands River.

This was the first that they had heard of the investment. Colonel Hore, with some 400 colonials, was holding out against a force of 3,000 Boers under de la Rey and had been besieged for nearly a fortnight. 20 miles were covered before halting at nightfall after vainly attempting to open up communication by Helio with the beleaguered garrison. Four Australians volunteered to try to get through with a message to Hore, and vanished into the darkness with the good wishes of the force. Before sunrise one of the quartet returned to report the complete success of their task, three of them having penetrated the *laager* before the enemy became aware of their presence. That morning, after a 10-mile ride, a little musketry play from departing and disgusted Boers, and the *laager* was reached. The relieved garrison took their release philosophically enough, without cheering or other effusion, merely rising in their trenches, pipe in mouth, and hands in pockets as they watched the relief file past. Yet these colonials, Australians and South African, had had a hard enough time of it during the past fortnight. They had *laagered* on a small sloping ridge, rising from a wide, but sparsely brush clad, valley, commanded on all sides, at 2,000-6,000yds, by higher hills. Here they held out with the carcasses of some thousand cattle and horses, killed mainly during the first two days' siege and dragged out under cover of darkness some 100yds or so outside and below the *laager.* Over 4,000 shells were fired into the *laager* during the fortnight as well as a continuous, if long range, musketry fire. Owing to the gallant defence they were offered, at the second day, safe escort to the nearest fort with the Boers retaining rations, boots and equipment. The offer was curtly refused. Before the siege was raised the garrison had lost a quarter of their number in casualties.

Metheun, who had abandoned further pursuit of the scattered Boers and was now on his way to Mafeking to refit, took over the beleaguered garrison. Meanwhile, Kitchener's force, after a day's rest, started back for Pretoria. While the main body moved directly, Colonel Legge, with the Australians and 'D' Company attached, moved in a more dilatory manner searching farms in the neighbourhood of Commando Nek where Steyn and de Wet were rumoured to be in hiding. Boers were encountered, shots were exchanged, but no serious fighting took place and Pretoria was eventually reached on 28 August. Two months' hard trekking had much depleted the horses in the ranks of the mounted corps. Consequently, when Ridley and Broadwood again moved out after a week's rest, many of De Lisle's II MI Corps were still unmounted. De Lisle himself was by now sufficiently recovered to take over command. The dismounted numbers of the corps, still numbering nearly one third, were left in the charge of Captain Robert Gordon, to follow later as soon as horses were obtained. Every company of the 6th was thus split into two, a separation that was destined to last a full month, the first and only time that the 6th MI was split into detachments during its three-year regimental history.

Lieutenant Clowes moved out in command of the mounted men of 'D' Company remaining with regimental headquarters. The first trek out was a short one, lasting only a fortnight. The operations were chiefly occupied in an attempt to clear the Magaliesberg in conjunction with Clements' infantry division. On 10 September a combined sweep of the crest of the ridge took place ending in a pitch fight at Zeekoehock where the 1st Battalion had fought so gallantly, exactly two months earlier to a day. The results of the drive and action known as Doornkloof were 38 prisoners taken and 227 waggons captured together with a quantity of supplies.

A week later the corps were back in Pretoria with orders to entrain for Honing Spruit in the Northern Free State. On 21 September the corps moved to Heilbron area. Meanwhile, on 9 September, Captain Gordon, after drawing remounts, had been suddenly ordered in the opposite direction as escort to 1,200 remounts destined for Buller's force at Lydenburg! He had under him Lieutenant Anley, Regimental Adjutant (Essex); Lieutenant Ferrar (Welsh) and Lieutenant Harrison (Argyll and Sutherland Militia) attached to 'D' Company. He had, in addition, odd details of South Australians and 150 New South Wales Mounted Rifles. With the exception of a stampede of the remounts at Oliphants River, when the mob galloped 8 miles into the *veld* before they could be stopped, the march to Machadodorp was uneventful and the animals were handed over to Buller's Remount Depot on 21 September. Instead of now being allowed to return to their corps. Gordon's little force was attached to the Machado garrison for mounted patrol duties.

The end of September saw the change over from regular to guerrilla warfare. With the British advance to the Portuguese border there was every expectation of the campaign being virtually over. The Boer armies in the field had been defeated and dispersed. Every portion of the country except the Bushveld of the far north had been visited and every petty town entered, if not permanently occupied. President Kruger was in exile, President Steyn a fugitive and only a few roving bands had still to be accounted for. Yet it was precisely from this date that the Boer was to show his true genius as a fighter. While Steyn started to reorganise the Transvaal, de Wet sent the Fiery Cross through the Free State.

With the passage from regular to guerrilla warfare the mounted troops, and not least the MI and colonials, were to come into their own. The infantry, from playing the major role, were to pass to the minor sphere of garrison duty and escort to convoys. A good deal of reorganisation took place; regular formations grew more elastic, expanding and contracting for ever varying requirements. Consequently, the term 'column' came into general use while the designation 'brigade' and 'corps' gradually dropped out of official reports. The first contingent of Australians, enlisted for one year's service, were also due for repatriation. Captain Gordon did not stay long at Machadodorp, though the next fortnight was to afford the only fighting in their three-month detachment. During this time, however, besides daily scraps on patrol, they took part in a short jaunt to the Carolina Valley. On 9 October they were ordered to Barberton to take over New Zealand remounts and patrol duties from New Zealanders going home. That night they gave a great farewell concert to their New South Wales comrades of the past year, also due to return home, and next morning left for their new station by train. The five weeks' picnic at Barberton were the quietest in the annals of the 6th MI, but even picnics weary, especially when accompanied by much horse sickness, and 24 November saw them on their way to Pretoria with no regrets at departure. After picking up remounts they rejoined their own column, in from trek, at Kroonstadt on 9 December, just in time for the next great adventure.

We left the II Mounted Infantry Corps on 21 September at Heilbron, which was to form the supply centre for General Dalgety's Colonial Division and other small columns operating in broken hill country of that portion of the Free State. On 1 October, Elands Kop, an isolated hill northwest of Lindley, was surrounded after an arduous night march in an endeavour to surprise de Wet. A sharp little fight ensued resulting in a few prisoners and some slight British casualties. The surprise was yet to come, for that evening at 10.00 pm came the news that de Wet had crossed the line in broad daylight with 2,000 men and 16 guns near Wolverhoek Junction and was now moving on the Transvaal. The news was correct enough, but his force consisted of only 800 men and half a dozen guns. The artillery was later to fall into the hands of the pursuit. De Wet had 24 hours' start, and the pursuit though taking up immediately was carried out with considerable caution and touch was speedily lost. Though it was certain that de Wet had taken the road to Vredefort, a week was occupied in covering the intervening 70 miles. After arriving at Verdefort the Colonial Division, to which II MI Corps was attached, spent a fortnight repeating the dilatory tactics of six weeks before, while rumours of de Wet's swelling force grew daily stronger. With the approach of the Colonial Division to Vredefort, de Wet had made an ostentatious break back through Schoemans Drift southwards to the Rhenoster River, then quietly doubling back, had crossed the Vaal in the open country to the west and joined Liebenberg with 700 Transvaalers awaiting his arrival in the Transvaal.

On 20 October, de Wet gave precise but disconcerting news of his whereabouts by heavily attacking General Barton at Frederikstadt. As soon as the news was learned at headquarters, columns were directed to Barton's assistance. While Colonel Hicks was directed along the Johannesburg Klerksdorp railway, General Charles Knox was sent from Kroonstadt to take charge of the Free State operations. Available were the Colonial Mounted Division, now reduced to 600 men by the departure of 400 Australians; De Lisle's II Mounted Infantry Corps, at present only 300 strong; and Le Gallais' VII Corps, some 600 strong. Knox brought with him the Royal Scots as escort to the convoy. Knox lost no time in crossing the Vaal. Le Gallais was recalled from Bothaville and the combined forces crossed on 25 September, coming into immediate touch with the enemy and capturing prisoners of whom the II Corps accounted for fifteen. Leaving Le Gallais to hold the Tigerpoort Hills, which at this point commanded the drifts over the Vaal for 15 miles, Knox pushed on with the remainder of his force to Potchefstroom, which was reached at 9.00 am on 26 September. Here he learnt of Barton's relief by Hicks the day before. De Wet was reported to have headed for Lindeque Drift 10 miles east of the Tigerpoort hills. As soon as he had collected his information, Knox started at dawn on the 27th for the Rensburg Drift directing Le Gallais at the same time to cross by Venterskroon Drift below the eastern spurs of the Tigerpoort Hills and then work round southeast. Once again, de Wet had doubled, this time under the northern spurs of Tigerpoort and over the Rensburg Drift for which Knox's main body was now heading. The II Corps came on the heels of his rearguard at 3.00 pm and pushed them back, stubbornly fighting, till the drift was reached at 5.00 pm. As the pursuit came in sight of the river, the *laager* had just inspanned and the riverbed was full of Boers. Whilst the Australian vanguard dashed in at a gallop and cut off the Boer rearguard capturing a Krupp gun, 'U' Battery and the pompom shelled the drift and *laager.* Just at this moment Le Gallais, emerging from the high ground south of Venterskroon, headed the retiring burghers from Parys and the east. A big success seemed on the point of achievement when the lowering clouds, which had been gathering all day, burst into a heavy thunderstorm with torrential rain. The setting sun was now low on the horizon and in the confusion de Wet got clean away, but not without the loss of a second gun to Le Gallais' MI and twenty-five casualties in killed and prisoners, including Commandant Wessels and four of Theron's famous scouts.

With water pouring off the rocky hills, the river quickly rose in spate. The crossing of the main body was much delayed, and the transport was not all over the drift until late on the afternoon of 28 September. The storm raged all night, rendering any thought of immediate pursuit impossible. Once again, this time through no fault of their own, the column lost touch with de Wet. Knox moved his troops to the vicinity of the line, there to start a drive westward on a broad front. Maxwell, with the colonials, was sent to Groot *vlei* as right column; De Lisle to Kopjes Station as centre; and Le Gallais with the VII Corps MI to Honing Spruit on the left. Knox marched with De Lisle. The drive started on 2 November. While Maxwell was to take the Witkopje hills round Vredefort, De Lisle and Le Gallais were to work in close cooperation north and south of Rhenoster Kop respectively and on to the junction of the Valsch River and the Vaal. The third day out contact was established with the Boer vedettes at Groot Pan, 10 miles east of Bothaville, which Le Gallais reached an hour ahead of De Lisle. A reconnaissance pushed into the town by Le Gallais that afternoon came under fire of the Boer guns south of the river, which the Boers were holding. An attempt to cross the river was frustrated and darkness put an end to the fight. Le Gallais bivouacked his force that night in the market square and sent to inform Knox, 10 miles to the north at Elandsvlei, of the situation.

Le Gallais' men were again in the saddle at 4.00 am on the morning of 6 November following the track of the Boer guns which led south from the town. Two miles further on Major Kenneth Lean, in charge of the 5th MI advanced guard, stumbled on a Boer picquet asleep and captured them without firing a shot. Realising that the *laager* must be close by, he cantered quietly forward with his company of seventy-six men to some rising ground in front. On topping the rise the *laager* suddenly came into view not 300yds distant, the men all asleep; waggons and guns outspanned and horses grazing loose. It was now 5.30 am, with dawn breaking, sufficiently light to shoot. Lean, who had previously sent a hurried message for the guns, dismounted at once and open magazine fire. The surprise was complete. Catching the first horse that came to hand and mounted bare-backed with only a bridle and rifle de Wet, Steyn and the majority of burghers galloped for safety into the blue abandoning the *laager* to its fate. Many horses, however, had stampeded, some 130 burghers found themselves without a mount. These pluckily decided to make a stand to try to save the guns. With this end in view they lined the stone walls of a garden belonging to a farm nearby and immediately set about covering the guns of the *laager*. For the next four hours a most determined fight took place, most of which has no part in the story of the 6th MI. Le Gallais, after sending off a report to Knox with De Lisle, came up to personally direct the fight from the farm.

Here follows a short account by Lieutenant-Colonel H.K. Umfreville DSO (Lieutenant, West Riding Regiment, at the time) of what occurred:

> 'As we charged, I was riding with my incomparable (for this kind of work) Australian bushmen. Watty Ross, our CO, dashed past me on his black Arab, miles the fastest animal in the column, and shouted to me, "I am going to get de Wet by his shirt tails." Le Gallais was close behind him, also Williams, Ross's Staff officer; Trooper Murphy, who carried Le Gallais' yellow Pennon, and I think two grooms, were all in a bunch together. Their objective was the farmhouse, which, with some outbuildings, formed Bothaville. Ross, Le Gallais and their staff actually rushed the house where de Wet was sleeping, and Ross burst into the room as de Wet, literally in his shirt, went out of the window and made his escape. His bodyguard covered his retreat and Ross had his lower jaw blown off; Le Gallais and Williams were both instantly killed, and only the presence of mind of Trooper Murphy saved the remainder, as de Wet's escort were firing through an open door at the end of a short passage which exposed anybody entering the house to view. Murphy closed the door and saved many lives by so doing. The shooting here was all a point-blank range, the Boer escort sheltering in a stone cattle *kraal* only a score of yards from the door.'

The command now devolved upon Major Taylor of 'U' Battery Royal Horse Artillery, who sent a further message for immediate help. By this time many

Boers from outside had recovered their wits and were returning to assist their comrades in the *laager* by crossfire from outside.

In consequence of the news the night before, De Lisle's corps were already saddled by 5.00 am and were on the move. Shortly afterwards the sound of Le Gallais' guns were heard in action and De Lisle at once galloped forward at top speed followed by every man who could hold the pace. By 6.30 am the intervening distance had been covered and the reinforcements had reached the drift from the town. Here Taylor's second message was received by mounted orderly. Knox, who had accompanied the ride, directed De Lisle to take charge of the fight whilst he supervised the dispatch of reinforcements as they arrived. De Lisle first sent up the VII MI Corps Reserve consisting of portions of the VII and VIII MI. These he directed to work round the right flank of the enveloping Boer counterattack and, as they were driven in, to finally take the Boer defenders of the garden in the rear.

The VII Corps Reserve was reinforced by New South Wales Mounted Rifles, the first unit to arrive, as soon as they had formed up. Next in order came Kitchener's Horse and the 6th MI who were thrown in on the right of the New South Wales Mounted Rifles in immediate support of the VII Corps troops grouped round the farm. De Lisle's gallop had been infectious and among the VI MI Corps were Lieutenant Derry, and 'B' Company who should have been on baggage guard but thought in so rollicking a fight, the baggage at Bothaville could look after itself. After the action was over this misguided enthusiasm was met with a full and merited rebuke! Finally, Knox sent up the West Australian Mounted Rifles Reserve. With these De Lisle reinforced Lean and the 5th MI on the west. It was now 9.30 am and De Lisle, with all available forces in position, determined on the final assault. For this purpose, he chose the 5th MI under Major Lean with the West Australians just arrived. The flash of steel on the rifles was, however, enough for the Boers in the *laager* who immediately hoisted a white flag and held up their hands in surrender to the answering shouts of the oncoming MI. The immediate gains in the victory were considerable, but there was no further pursuit. Five guns of various calibre and one Maxim were among the spoils, together with all the Boer waggons, a large supply of gun and small arms ammunition and a great quantity of saddlery and clothing. Seventeen Boers were buried on the field and prisoners amounted to seventeen wounded and ninety-seven unwounded men. Le Gallais died at 8.30 pm in the evening, universally mourned by both columns with whom he was a most popular and trusted leader. The British losses included thirty-eight killed and wounded, of whom ten were officers. Bothaville was essentially a mounted rifleman's battle and continued to be a source of pride and inspiration to that branch of the service for the remainder of war. The chief honours from the fighting belonged to Le Gallais and his VII Corps , the account of which, very much abridged, has been mainly taken from *The Times History of the South African War.*

Then followed a well deserved week's rest at Kroonstadt. Here a draft of sixty NCOs and men under Lieutenant A. Cameron from the 1st Battalion which left Belfast on 27 October was awaiting their arrival. In the middle of November the little column moved out to Rhenoster Kop and a big isolated hill 20 miles north west of Kroonstadt. Here headquarters was established and, working in conjunction with Maxwell's colonials, the whole country between the Valsch and the Vaal was swept clear of cattle and the Boers were kept on the move.

These operations did not go unchallenged and, from time to time, there was stiff fighting, especially around Vredefort where a mass of low wooded *kopje*s on both sides of the Vaal gave excellent cover for guerrilla warfare. Kopje Aleen, a big mass of wooded hills nearly 20 miles long and 10 miles deep with a mass of ravines running down, which lies south of the Valsch and some 30 miles south of Kroonstadt, was also visited. After three weeks' outing they returned to Kroonstadt with the captured stock to refit. Here on 9 December, Captain Gordon rejoined with his detachment, as previously narrated. After only one day's halt in the town, De Lisle was off again, this time working with two other small columns on a 20-mile front. The trek took them to within touch of Reitz and the east, then back to the line with captured stock, and again south to Ventersburg, which was reached on 17 December. From Ventersburg they were to have moved on Winburg, when sudden orders came to move at once to Ventersburg Road on the railway there to entrain. Over 150 miles had been covered in the week but each day's work must have averaged over 30 miles, by the time every farm had been visited and all available cattle rounded up. Entraining took place immediately on arrival on 18 December and, travelling all next day, Hanover Road was reached at 5.00 am in the morning of 20 December, the starting point of the next four months' adventure.

De Wet had turned the tables with a vengeance and had carried the war into British Cape Colony. The immediate cause of this move was the invasion of Cape Colony by two of de Wet's lieutenants. On 16 December, Commandant Hertzog had crossed the Orange River at Zand Drift, 30 miles north of Colesberg, with 1,000 men. Commandant Kritzinger had, on the same day, crossed at Odendal Drift 85 miles to the eastward with 700 men. As soon as Lord Kitchener realised that trouble was threatening in Cape Colony, he sent Major General Settle to take over the direction of the operations with headquarters at Hanover Road, halfway on the branch line connecting the two South African Railway systems through Cape Colony at De Aar and Naauwpoort Junctions respectively. Hanover Road, then, was De Lisle's destination, where a concentration of various Free State columns was taking place. Hertzog's ultimate destination was Lambert's Bay in the Clanwilliam district, where European sympathisers had promised a ship loaded with munitions of war and volunteer sympathisers in the Boer cause. Reinforced by these practical expressions of goodwill Hertzog hoped to swell his force with a big rebel contingent and hold to ransom the rich provinces around Cape Town. It was a bold and romantic conception which entirely overlooked the fact that Britannia was still undisputed mistress of the seas!

Hertzog, after occupying Phillipstown on 19 December, threw out feelers for his further line of advance. Finding Settle barred his road south, he crossed the line on the night of 21 December at Houtkraal on the De Aar-Kimberly line and, moving due west, marched on Britstown, which he occupied on 23 December. Meanwhile, De Lisle had arrived at Hanover Dorp where all was found quiet and in order. The next day the column moved off at 5.30 am, marching north with Colonel Parson's column co-operating. There had been very heavy rains and the country was waterlogged, much impeding the transport. Hertzog, in forcing the line at De Aar, had not only engaged an armoured train but had also captured four Yeomanry companies incautiously sent out from De Aar to engage him. The column was consequently sent to De Aar to straighten the situation. De Aar was reached on 23 Decmber when it was learned that Hertzog had moved on to Britstown. Thorneycroft's column was ahead, and entered Britstown on 24 December on the heels of Hertzog. De Lisle, following, entered on Christmas Day and was fortunate enough to find a sumptuous Christmas dinner, prepared by Dutch sympathisers of Hertzog's commando, awaiting their arrival. Under the circumstances the inhabitants were not sorry to transfer their hospitality and allegiance since Hertzog had commandeered all their horses and £2,000 of goods on departure!

On Boxing Day the column caught up with the commando some 12 miles north of Britstown. The Boers were holding a long ridge covering the British advance over an open plain, with their right flank resting on the Ongar River and a group of high *kopje*s protecting their left. Skilfully hidden in the thick *karroo* scrub, the scouts almost walked into the first line of riflemen at the foot of the ridge before they were discovered. On being ejected from there they scrambled up the steep-sloping ridge face and, lining the summit, opened withering musketry fire on the column wheeling into action. The British artillery unlimbered and commenced firing, sweeping the whole ridge from end to end with shells. Under cover of this barrage the column, with the Australians leading, swept forward to the attack. The Boers held staunchly to their position and the crest was not gained till three hours later, when a wild rabble of Boer horsemen were seen scattering over the plain in the rear. The gunners hastened their departure while the Australians rode three miles in pursuit. In their flight the Boers abandoned a number of waggons, besides leaving five dead, seven wounded and several prisoners in British hands. British casualties were three killed and four wounded, among the latter being Colonel Knight for the Australians who, while gallantly leading his men, was shot point blank at 40yds by a mounted Boer. This assailant paid dearly for his exploits for he was brought down with seven bullets in him by the enraged colonials, shooting from horseback. The force then crossed the river to its western bank and bivouacked for the night. On one of the captured waggons were several barrels of brandy, while on another the remains of the Prieska garrison's Christmas beer, captured while proceeding to that place from De Aar. Littering the Boer main position, as well as on the camp site by the river, were hundreds of empty liquor bottles commandeered at other people's expense, which showed the Boers had not turned down a little Dutch courage to aid them in their Christmas cheer. After all, for the past six months they had held a long preliminary fast.

Thorneycroft's column and Parkes' Irish Yeomanry now joined the column and the three columns trekked together till 6.00 pm when De Lisle separated and, marching south for another 12 miles, eventually halted at 11.00 pm. The halt was merely a bivouac without fires for a little fitful sleep and the march was again resumed in the early hours of the morning. Travelling south west, touch was not regained for several miles but, as daylight broke, the Boers were located at the tiny hamlet of Vosburg and driven out without difficulty. On leaving the hamlet a Cape cart, drawn by two horses, was seen hurrying off with all possible speed. Scenting an important capture the Gordon company dashed in pursuit only to find it contained two Dutch farmers with their wives and children, who had come in for a morning's shopping and were now escaping from the additional excitement to the days' outing. Carnarvon, with only a weak garrison of Yeomanry, appeared to be the Boer objective, but now the carcasses of worn-out Boer horses left a tell-tale trail easily taken up and the pursuit was too closely followed for any risks to be taken with a British garrison, however small. Trekking day and night, with only short intervals for rest, Carnarvon was entered on the 30th; the Boers themselves, leaving the town on their right. The entry of the advance guard coincided with that of another body of horsemen entering from the south. These proved to be Yeomanry returning from reconnaissance who immediately took cover and it was a quarter of an hour before identity was established, luckily without exchange of shots, as was often the case. At Carnarvon the column exchanged their worn-out mounts for the fresher horses of the Yeomanry and, at midday, resumed the chase. Many dismounted men had perforce to be left behind. Though the Boers were able to commandeer remounts ahead of the British, their horses too were feeling the strain and many dead horses marked the road they had taken.

Fraserburg was the next Boer objective, lying some 60 miles to the southwest. On 3 January 1901, at the Zak River the chase slackened owing to the shortage of provisions, for the Boers were ahead commandeering flour and forage as well as horses. Kitchener had previously abstracted five of the convoy waggons with their supplies for a new cavalry column forming just as the force was leaving De Aar. A move over the Niewveld mountains towards the railway was consequently to meet a fresh convoy. Thorneycroft's column moving into Victoria West for the same purpose. All three met in the rugged passes of the Niewveld range and the columns turned northwards again towards Fraserburg 40 miles northwest. The second day out, however, they were recalled to Beaufort West on the railway to entrain.

With Hertzog safely shepherded into the rugged and rather barren country of the Western Colony it had been decided to forestall him in the south and drive him into the sparsely occupied wilds of Namaqualand. Before entraining, 'D' Company under Gordon were sent on a two-day commandeering expedition for fresh horses. A number of farms in the glens of the Niewveld range were visited and a considerable number of useful Cape horses were collected. In most cases the farmers were glad to sell before the Boers had a chance of commandeering. In recalcitrant cases the threat of report and a £5 reduction in the British IOU brought the men to their senses. Occasionally an enraged cock ostrich, no mean opponent, gave excitement to the quest. Returning with their remounts, the company halted for the night at the small hamlet of Rhenoster Kop on the north of the range some 15 miles from Beaufort West. That evening they saw a convoy which halted some distance off without coming into contact. Next morning, at 3.00 am, they were awakened by shouts of 'hands up' and 'move and we shoot' in unmistakable British voices, while helmeted and khaki-clad figures excitedly flourished Lee Metfords in their faces. Their captors were the newly raised Railway Pioneer Battalion, fresh out from home, and explanations given in a strong Scottish accent only confirmed previous suspicions. It was not until sometime after, when an officer had been brought and had been confronted by Gordon, that the situation was eventually cleared. In explanation it may be stated that in a composite regiment, while 'B' Company often talked together and sang in Welsh, 'D' Company had reverted to the Doric of the countryside seldom heard in its purity in a regular unit. De Lisle, in fact, often insisted that an excited orderly from 'D' Company must be accompanied by an interpreter!

From Beaufort the column was carried to Picquetberg Road when it detrained on 12 January, marching out at 5.00 am. and halting 14 miles out on the road to Picquetberg. Picquetberg, perched on a mountain slope, overlooking a fertile and well-cultivated valley, was entered on 14 January. Here they got a great reception, being regaled to a fine spread on tables covered with white cloths. For the first time in all their wanderings they were among friendly population. Scarcity of water for the requirements of so many animals forced the column to move onto a small hamlet six miles distant. Here a week was spent in pouring and continuous rain while preparations for the further advance were being made. During the week's halt communication was kept with Lamberts Bay, which was now a British base of supply, by means of strongly escorted convoys.

Hertzog, meanwhile, having shaken off the pursuit had profited by his three weeks' grace, hurried on westward to the sea. Pausing at Calvinia for a few days' wild carrousel, he moved on towards Clanwilliam, first sending a small commando to Lamberts Bay to meet the promised ship. On arrival they found their expectations fully realised. The ship was there all right, loaded with ammunition and a fine contingent of fighting men, only it flew the white Ensign and delivered its munitions in a salvo of shells to the Boers waiting expectantly on the beach. This was rather a blow to hopes of further success. Clanwilliam, however, still lay unoccupied, ready for plunder before the weary trek back was resumed. The road from Clanwilliam to Calvinia lies over a spur of the Pakhuis mountains. Once the Boers were in possession of the pass, they would command the town. A party of West Province Mounted Rifles, one of the numerous local corps raised for the garrisoning and defence of the colony, had been sent to seize and hold the pass. On 23 January these had come into contact with the Boers.

Preparations completed, De Lisle moved north on 21 January, through the Picquetberg mountains by Greys Pass and onto Modderfontein on the Olifants River. Since this was very swollen with the late, heavy rains and the water high over the bridge, the crossing was, in consequence, much delayed. Moving up the Oliphants valley through magnificent scenery, Clanwilliam was reached at 10.30 am, just as the Boers were engaging the Cape Mounted Rifles. From Olifants River, however, 'D' Company had been again dispatched on a two-day horse commandeering expedition, guided by a Hebridean Scot called MacLachlan, from Lamberts Bay. The country was wild and rugged in the extreme, the mountain sides covered with huge rhododendrons and other flowering shrubs down to the riverbank, very different country to the Free State uplands. Numerous snakes darted from the undergrowth. Gordon, who had not lived among snakes all his life for nothing, jumped from his saddle and seizing a 10ft serpent by the tail and, pulling it rapidly towards him, crushed its head beneath his heel. Again, another big snake was caught but this time the serpent rose to strike, then turned to escape as he leapt aside. Twice the manoeuvre was repeated till Gordon got a favourable opportunity to seize its tail when immediately he gave it a crack as with a stock whip and with the crack the head flew off!

At farms visited the people were friendly and at one farm a woman, seeing the party approach, mistook them for Boers and promptly fainted. That night at the farm, at which they slept, the people, after an excellent supper, regaled them with English, Scots and Irish songs while Gordon played the accompaniment on the piano. Returning next day with their purchases down the valley the company found 80yds of raging torrent between them and the road the column had taken. Determined to reach Clanwilliam that night Gordon put his horse at the river and after a severe struggle reached the opposite bank. His horse, however, fell into the water before landing could be made and swam back to the opposite bank. Gordon then stripped and swam over, arriving somewhat exhausted. Again riding his horse, he swam the stream again for his clothes and, after dressing, remounted and rode down for a ford. It was clearly impossible for the company and remounts to cross here. Six miles lower down some shallows were struck and tall trees had been thrown over the deepest channels. Over these perilous bridges, a foot under water, the company crossed with the assistance of a local native. The commandeered remounts, however, stuck on a sandbank and refused to move. It took two hours of cajoling till, with the assistance of the Teamsters of a solitary waggon which at last hove in sight, the mob was got over just before dark. At Clanwilliam the company learnt of the death of Queen Victoria. Both Dutch and English joined in paying sincere respects to her memory. Here a week's halt was made while supplies could be fetched from Lamberts Bay. These were not to pass unmolested and 'A' and 'B' Companies forming the escort were heavily engaged by hungry Boers who also wanted to replenish an empty commissariat.

On 28 January General Bethune arrived with his column and, as senior officer, assumed command of the operations. Next day the force moved out over the Pakhuis Pass to engage the Boers and start the long drive back to the Free State. The road was terribly steep and only the summit of the pass was reached by nightfall. The MI did not reach camp until 10.30 pm Early the following morning the descent was begun to the Doorn River. The country was incredibly rough; great boulders loomed everywhere with steep nullahs constantly crossing the winding road. In the latter were the still smouldering campfires of the retreating Boers. A few miles from the foot of the pass the Doorn River was almost reached when the Boers were discovered occupying several suitable *kopjes* on either side of the road leading to the drift. The strong natural position and numerical superiority of the Boers forced De Lisle to halt and send back for reinforcements before attacking next day. In the evening a small patrol under Captain Gordon, with Lieutenant Clowes and an orderly, went forward to find out if certain ravines were practical for horses. Pushing 2 miles ahead of the outpost, which was further than intended, they fell into an ambush of the Boer picquets and were all shot at close quarters. Gordon was dismounted and was peering in the bushes for evidence of tracks when the attack came. Shot in the foot he had some difficulty mounting and as he galloped back amid a hail of bullets, he passed the bodies of Clowes and his horse, both shot dead. On reaching camp there was no trace of the missing men and an ambulance was sent from camp to meet later Private Clarke, dangerously wounded in the groin, staggering back to camp. He had been treated with great brutality by the Boers. He died of his wounds before he could be brought to Clanwilliam. Some of the enemy came forward under white flag to assist in looking for Clowes, but the body could not be found that night. Next morning when the combined forces deployed for attack the position was found empty. Clowes' body was discovered some distance from where he fell, rifled of everything worth taking. He was buried under a cairn by the roadside, De Lisle reading in the burial service before the column moved on.

With 'Boomerang' Gordon a casualty in hospital this ends the graphic diary of his batman, Charles Gordon, which has so far provided the backbone of the story. Cameron took over temporary command of the company but as he was the only officer, Harrison having been sent back with details from Carnarvon and later developed enteric fever, Lieutenant Strong of the Bedfords was soon posted to command, a most popular and able officer. Both were killed on 6 June at Graspan in the Free State.

De Lisle now received orders to alter direction and swing to the south along the crest of the Zwartbergen and down into the valley of the Doorn River. The country was very wild and rugged with steep slopes covered with thorn scrub. Progress was necessarily very slow. High precipitous cliffs towered everywhere above the road. In the descent into the valley five waggons were overturned. The road itself wound round the face of the mountains, broken everywhere by precipices and in places roughly cut out of the rocky face of the cliffs themselves. The ford over the Doorn River proved almost impassable and a halt had to be made for repair. Reascending from the valley only the guns and a few waggons reached the top of the pass that night. All that night until noon on 1 February men were working, loading and offloading waggons and bringing them up with double teams. The loads, meanwhile, being carried by pack horse or manhandled. Not till 4.00 am on 2 February did the last waggon reach the summit and only two and a half miles had been covered in 38 hours! It was the worst day's march struck in the campaign. Camp was made at Elands *vlei* where a tributary of the Doorn comes down from the Roggenveld mountains. In all, five days had been spent in covering 45 miles between Clanwilliam and the Doorn. The next two days' march led over 50 miles of open *karroo veld* and camp on 5 February was made under the Roggenveld mountains, some miles from the Boshkloof Pass behind which lay Calvinia. The Boers were holding the pass. At 11.00 pm, the New South Wales Rifles moved off to turn the pass from the north and cut off the Boer retreat; the main column moved at 3.00 am. As they reached the foot of the pass at dawn the Australians came into view crowning the heights 2,000ft above. They lost several of their horses over the precipice in this fine performance. Once through the pass the town came into view some two miles distant. De Lisle thereupon extended the 6th MI on a 3-mile frontage with lieutenant Derry and 'B' Company as advanced guard. Advancing at a canter which quickened into a gallop, a company was sent to each flank encircling the town while Lieutenant Derry led his company at a gallop through the main street. The town, however, was found to be evacuated and a great welcome awaited the troops from the loyalists among the population. Hertzog's two lieutenants, Brand and Pretorius, had behaved very badly during their visit, flogging the well-disposed natives, insulting the Queen's memory, fining loyalists and smashing up the English church. It is but fair to add, their conduct was most unusual, due probably to the bitter feelings at the failure of the great raid.

Next day 7 February, Scobell's force came in from the north during the afternoon. Three long and weary marches followed to Williston, the last 72 miles being covered in 48 hours on the heels of the disappearing Boer convoy, which did not stay to fight. Here orders were received to concentrate at Carnarvon, which was reached on 16 February after four long and tiring marches over the Zak River through the many valleys and rivers of the scattered Frazerburg range of hills. Fighting, however, was over for the moment. It was now just a year since the regiment had been formed, during which period some 7,000 miles had been covered in the saddle; no bad record of continuous hard riding!

16 February marked a turning in the campaign in Cape Colony. On 10 February de Wet, encouraged by the optimistic report of his two lieutenants, had crossed the Orange with what reinforcements he could collect. Invasion, however, had lost the glamour of a year previously and only 2,200 burghers, somewhat reluctantly, responded to the call. Of these, 800 turned back faint-heartedly when the Orange was reached leaving only 1,400 stalwarts as a forlorn hope. Immediately on crossing he found his way south barred by waiting columns and, like Hertzog, was forced to take the road to the less populous west. Handicapped by adverse weather conditions and constantly harried by pursuing columns, and having summoned Hertzog to join him at Prieska in the north of the colony, his sodden and dispirited force were already ripe for a return to comparative safety across the border. Moreover, a close acquaintance with rebellion had brought disillusion to all.

While in the west, if the Freestaters had proved overbearing, the colonial Dutch had contributed aid little more substantial than good wishes and Cape brandy. As Hertzog swung north, other columns from the line took up the chase where Bethune had left it. When, therefore, Hertzog and de Wet did eventually join forces at Strydenberg in the Hopetown District, on 22 February, the great adventure had already failed and the thought uppermost in the minds of each was how best to escape back into the Free State. Here then we may well leave them.

After halting at Carnarvon to refit, the march back to the line via Britstown was made in more leisurely fashion. De Aar was reached on 2 March and entraining took place that afternoon, this time at Cookhouse on the railway, 80 miles north of Port Elizabeth and in the Bedford District. The same day saw a draft of thirty NCOs and men under Captain S.G. Craufurd and Lieutenants White and Mitford leave Belfast for the 6th Mounted Infantry. After detraining at Cookhouse the force marched to Somerset East where they were entertained by the townsfolk with cakes and tea.

Kritzinger, *beau sabreur* of the cause with his lavender kid gloves, gold mounted riding cane and polished polo boots, had penetrated to the hills above Port Elizabeth in the Eastern Provinces. But here the comparatively large English farming community acted as a salutary check upon active sedition.

A slight sketch of Kritzinger's achievements to date may here be given. After crossing the Orange in the middle of December he had made straight for the Zuurberg; a range of hills which lie to the north of the railway connecting Stormberg with Rosemead Junction on the Naauwpoort line. From there he moved southwest to Aberdeen and Willowmore, or only 50 miles from the southern coast. Early in February he broke back, always with a group of columns under Douglas Haig on his heels. At this juncture every available column was called for the great de Wet hunt further north. Finding the pressure eased and with only Gorringe and the colonials to watch him, he determined to stay on in the colony and transferred his activities to the block of country lying between Port Elizabeth and East London lines, where we take up the story.

Moving out from Somerset East on the morning of 6 March, Kritzinger was found in position near Pearstown and a running fight over very difficult country took place all day. Bivouac was pitched in the dark. Kritzinger retired north into the Bankberg Hills and the fight was resumed early next morning, Gorringe cooperating. A Boer patrol of nine men was successfully ambushed and camp eventually pitched late in the afternoon. Kitzinger, who had no particular objective and was out merely to annoy, now moved east to the Little Fish River and turned south in the direction of Grahamstown, still closely pursued. On 12 March the drifts over the Great Fish River commanding the town were reached, which Captain Craufurd with his draft and small force of Colonial Defence irregulars was found to be holding. Captain Craufurd had reached Grahamstown on 6 March, there to receive remounts and await De Lisle's arrival. A bivouac was chosen on a wide-open space near the biggest house in the town. Sending up to ask permission to use an empty barn, Craufurd was met by Mr Graham the proprietor and grandson of the founder of the town. He at once set about finding comfortable billets for the men with a plentiful supply of straw and insisted on the officers coming up to the house. After dinner Mrs Graham apologised for there being no drawing room as it had been turned into the guest bedroom! She explained this gave the officers a private entry and that the adjoining conservatory formed a splendid bathroom as it had a fine, deep ornamental tank. Part was screened off as an office. During the intervals of Riding School, Mr Graham placed himself and his trap at the officers' disposal.

Three days later moving off, in addition to a splendid hamper for the officers, every man was provided with a pie, cake, fruit, beer, and cigarettes for his wallet. A friendly correspondence was carried for the next year or two. On moving off everything had to be requisitioned, cooking pots, mugs, provisions and bread, because those drafted had nothing but what they stood up in. A few old hands, as details, had joined up but the men drafted had only had two days' riding in their lives. Once clear of the town Craufurd quickened the pace to a sharp trot, halting from time to time to pick up his fallen forces, a drastic riding school but the men might be called on to fight mounted at the drifts 12 miles further on. Mounted patrol duty with English farmers of the defence force and old hands took the place of Riding School on 11 March and when De Lisle arrived on 12 March Craufurd was able to hand over his draft, sore but able to stick on. It was just as well, as a 21-mile ride to Adelaide followed next day. The Boers had passed both sides of the town without entering the day previously. From Adelaide, De Lisle moved north through the Winterberg to Tarkastadt, always in desultory touch with the Boers. From Tarkastadt, where the waggons were left behind, the force moved light in torrential rain to a group of hills 15 miles north of the town known as Palmistfontein. Here, on St. Patricks Day, De Lisle took up a position 8 miles long with only 400 men, there to await co-operation with Gorringe against the Boers in that position. Gorringe, however, was late and De Lisle with his gallant 400, two pompoms and two 12lb guns of 'B' Battery determined to attack Kritzinger in position with 600 men. With 6th MI on the left and the Irish Yeomanry on the right, the attack was launched. Both opposing *kopje*s were attacked in fine style and with great dash, captured and held all day under a hot fire. Kritzinger slipped away late in the afternoon through a gap left by the defence force who were holding the Elands River, thus avoiding almost certain capture. Gorringe, delayed by rains, arrived at 5.00 pm just too late to secure the prize.

The disappointed columns returned to its waggons at Tarkastadt that evening glad to get into dry and comfortable quarters. From Tarkastadt they entrained for the country north of the Zourberg on 21 March, detraining at Burghersdorp the day following. From here they patrolled the country to the north in case Kritzinger should try to break back over the Orange.

The following week passed pleasantly without much excitement. The column worked very wide so the flanks, often four and five miles out, had many opportunities for quiet halts. Generally, some farm was selected at about the column midday halt where pancakes would be cooked for the officers and stew provided for the men. At one farm, when a three hours' halt was known to be due, 'D' Company struck a lucky find. The owner's wife, an English lady with both husband and son at the front, asked if the halt would be long enough for lunch. Finding it was for three hours she immediately prepared hot baths for the officers and a fine lunch at which she and her two daughters insisted on waiting; protesting it was the only personal service they could give. Roast pork was prepared for the men with jam and duff, after which a visit was made to the orchard. Then off they rode with a cheery farewell.

The picnicking did not outlast the week for Kritzinger and his men, feeling their way for a trial break back. On 29 March De Lisle took the 6th MI and 50 Yeomanry without baggage to Bull Hoek where they bivouacked for the night. Bull Hoek was a curious cup in the hills, entered by a narrow pass and entirely surrounded by high hills. A regular little death trap, it had so far been avoided by friend and foe alike and rumour had it that sheer bravado had prompted the excursion. There was little sleep, for in spite of picquets on the hills all were rather *que vive.* At dawn, to everyone's relief, they climbed the encircling hills, glad to be out of the very picturesque little hollow, and were able to join in with the columns driving the Boers past the front. The British presence, however, caused a very wide detour as soon as discovered by Boer scouts, ranging far afield. The next few days were spent driving Van Rennen, one of Kritzinger's lieutenants, over the Orange with his commando of 300 men. The drifts should have been held and Van Reenen captured, for six columns were in position awaiting his crossing but, somehow, he got through. Kritzinger doubled back with most of his force, there to continue raiding.

Orders once more came to entrain and by 6 April De Lisle was back at Kroonstadt, this time under General Elliot, under whose orders the column continued to serve to the end of the war. From now on the 6th MI were no longer to have the sole honour of representing the Gordon Highlanders Regiment in that branch of the service in the field. Other contingents had been forming at home to take their share, a section in the 12th MI from the 2nd Battalion and one in the 19th, both operating in the Eastern Transvaal. While the 23rd MI from home brought a whole company to operate chiefly with the 9th and 15th MI in the Northern Free State and a further half company to serve mostly in the Western Transvaal. While other sections were formed in the later MI regiments whose records have now been lost.

Chapter 8

6th MI driving to plan - clearing the Northern Free State

On arrival at Kroonstadt, De Lisle's column found they were one of three operating under Lieutenant General Locke Elliott who had been entrusted with 6,000 mounted troops for general clearing operations in the Free State, north of Bloemfontein. The three columns were:

Bethune's cavalry brigade, reinforced by a new South African corps known as the Prince of Wales Light Horse.

Broadwood's cavalry brigade, now consisting of the 7th Dragoon Guards and three new regiments of Imperial Yeomanry.

De Lisle's column, of which the South Australians were more or less a new corps.

Thus, the division had many new hands among them.

The first sweep through the country was more or less experimental. The capture of stock, destruction of all supplies and escorting in of such refugees as were willing to come in, forming quite as big a part in the scheme of operations as actual engagements with the enemy. The scope of the sweep was the triangle between the Vereeninging Kroonstadt railway, the Rhenoster and the Vaal. The column moved out on 10 April 1901, De Lisle's column being on the outer or left flank. Another small column under Lieutenant Colonel Ingouville Williams was co-operating on the extreme right from Wolverhoek Station.

On 14 April, Major Sladen, Yorkshire Light Infantry, took over command of the battalion from Major Pennefather, who had been invalided from the Western Colony. There was very little fighting, but a vedette of the South Australians managed to get captured and stripped, their excuse being that they were at the time 'sitting and thinking of roast beef' instead of looking out for Boers. At Vredefort a party of Cape refugees wished to be brought in with their belongings as the country was being cleared of supplies. 'D' Company with some supplementary waggons and two squadrons of the new

Imperial Yeomanry from Broadwood's column attached were sent to do the work. Picquets were posted on the surrounding hills to prevent surprise but the Boer scouts contented themselves with watching, sending a message in under a white flag to say the actual evacuation would not be molested. The day was very hot, the refugees anxious that nothing should be left behind and the morning wore on. Finally, the captain and support half of the company lent a hand. It was thirsty work so, when all was finished, he asked for a drink. A cup of rather blue water, when water was seldom clear, was brought and before he could stop himself, he had swallowed the cup full. 'Good heavens, what's that?' he gasped as thoughts of poison flashed through his mind. 'The finest Cape Dop (Brandy) and many years old at that' answered his smiling hostess. On the way back when clear of the drift outside the town the Boers tried a few parting shots to hurry the departure. The dash and energy of the counterstroke led by the captain in person was as much a surprise to the burghers as to the bewildered Yeoman which he led.

A day or two later a squadron of Broadwood's new Yeomanry were ambushed near the Rhenoster River, one officer and thirty-five men being captured. They were later released after being relieved of their horses, rifles and such other accoutrements as the Boers had use for.

During this experimental drive one or more officers of the South Australians were attached to the advanced and rear-guard companies to see how scouting was carried out. On one occasion a young Australian officer turned up for such instruction on rearguard, having been with the same company on advanced guard a few days previously. 'I thought you were a Captain,' said the captain of the rearguard cheerily, on seeing his acquaintance arrive as a lieutenant. 'So I was,' answered the Australian, 'but my major reduced me!' Further enquiry showed the junior ranks were having quite a cheery time of promotion, the scale sliding up and down according to the breakfast complexion of the officer commanding. Unfortunately, a private report to De Lisle, on reaching bivouac, put a stop to the sporting sideshow in promising military careers. The line was again reached on 20 April, the ten days' outing resulting in the capture of 50,000 head of stock, seventy vehicles, and about 100 tons of flour and grain of which William's column attached accounted for about one third. How misleading these figures were to the real progress of the campaign is shown by the casualty list, both British and Boers having each three killed and five wounded.

The next sweep, however, was to be more ambitious and a little more complicated in design. It was in fact an attempt to sweep the far bigger triangle which had the same line of railway for base and Botha Pass as apex. Other columns were again called in to co-operate with Charles Knox from the south between Senekal and Lindley; Western, who had replaced Williams in command of his column, again in prolongation on the left; and with columns from Standerton and Heidelberg in the Transvaal to hold the drift over the Vaal. Elliott's main drive was to be in three places, taking the form of a rough S-shape. First eastwards on a 40 mile frontage between Heilbron and Lindley, then swinging north between Heilbron and Frankfort to the Vaal, then south again with the Klip River as his left flank. There, early drives differed from those of some months later in that there were no organised butts to drive against, while clearing the country of supplies occupied quite as much attention as the capture of prisoners. They were in fact more in the nature of great cavalry raids. Night marches were constantly undertaken to surprise *laagers* or farms and columns moved in close co-operation, but at night there were many gaps and avenues of escape, while daylight was chiefly occupied in clearing country of stock and food supplies. In official accounts no distinction has been made between the two operations. All these movements came under the general name of 'drives'. To those who took part, however, the methods employed varied so greatly that it may help the reader to follow the course of events more easily if 'raid' and 'drive' are each used in the proper place.

Elliot moved out on 24 April. Bethune on the left, De Lisle centre and Broadwood on the right. Boers were met the second day out, but no serious resistance was met till the swing north at Bachante on 2 May. From here, three days' foray was made on light kit. Broadwood and De Lisle's transport being placed together in the meantime. A small *laager* was captured yielding five prisoners, eleven vehicles and a good deal of stock. From Bachante, De Lisle was temporarily transferred to command Broadwood's column. Next morning the combined column found itself in touch with strong parties of Boers. Tafel Kop, covering Vrede, was held by 300 Boers and another commando, 250 strong, was working on the left front. Some three hours of skirmishing ensued.

Two active and enterprising young Boer commandants were in charge of the district with commandos composed mostly of young men like themselves. One, a young Scot by the name of Alec Rose, passes out of history with the close of the campaign; the other, Manie Botha, nephew of General Louis Botha, was to rise to high rank as a British General in East Africa in the Great War. During the following three months the column was to get on very close terms with these two fine young leaders who fought, in all, twelve separate engagements of three hours' duration during this period as well as innumerable skirmishes. The 6th MI and these two commandos had real respect for each other and, quite unknown to the authorities, from time to time, through medium of the regimental doctor attending Boer casualties by white flag request, a bottle of whiskey or tin of cigarettes would pass on the one hand, with samples of coffee substitute, or a platted straw tobacco pouch worked by the sisters of the men in the field, on the other. It was all very irregular; but doubtless tending to make the eventual settlement easier. The figures, though not, of course, the faces of the leaders, were well known to both sides by constant contact. For the next week Tafel Kop, a high, flat-topped hill some 2-3 miles long, was made headquarters while foraging parties ranged the country round.

On 5 May the picquets of 'C' Company were heavily attacked in a thick fog and the Boers eventually driven off with casualties, one shell killing three Boers in a neighbouring *donga*. Halfway through the week, camp was shifted to another side of the *kopje*. The grass of the old camp was very good and 'D' Company was left on grazing guard. It was a day of alternative fog and sunshine. The Boers knew the camp had moved and giving ample time, as they supposed, for the rearguard to move off, eventually crept up under cover of the mist to search the ground for cartridges etc, often casually left behind on striking camp. Hearing one or two short bursts of fire the captain, who was sitting below the picquet line reading a book, strolled up casually to see what was up. The mist had again temporarily fallen, and he was standing on the skyline, pipe in mouth, trying to take in the situation. Suddenly the mist lifted and for the first and only time in all their casual acquaintance he and Manie Botha found themselves face to face. Botha, armed, raised his rifle instinctively and fired point blank, then dropped to creep away. The bullet hit the bowl of the captain's pipe, broke the pipestem in his mouth, and knocked him head over heels with the shock. The picquet line seeing their captain fall, at once fixed bayonets and charged. The Boers, intent rather on loot than battle, hurriedly made off, losing three of their number wounded before they got clear. By now the sun had drawn off the mist and with the camp roused the Boers cleared off for good. That afternoon the Boers asked for the ambulance for the three wounded men and Manie asked if the captain was badly hit. Hearing what had happened he had the cheek to claim that he had only aimed at the pipe and sent a corncob pipe to replace it! The Boers had some reason for their activity. Vrede, their headquarters, was only 15 miles distant, and the little town had had to be hastily evacuated until the British should move on. Vrede itself was cleared on the 8th after a wearisome night march by 'D' Company, a squadron of South Australians and a gun, all under Major Sladen, but the Boers had already done most of the clearing. They had, as a matter of fact, fought another three hours' action doing most of the attacking the day previously whilst the womenfolk and other burghers were clearing out of the town.

The march north to the Vaal was resumed at 2.00 am on 10 May and, shortly after daylight, the Boers again attacked, and a sharp three-hour scrap followed. A small party of seven South Australians were cut off on the right flank and captured to be released later minus their horses, rifle and such equipment as the Boers needed. The Boers, however, lost one killed and four wounded in the morning fighting. Next day 'D' Company were left on the campsite to get in touch with Coville's Yeomanry, who were expected an hour or so later whilst the main body moved on. Finding only a small body of troops, the Yeomanry, who were newly out from home, imagined them to be the Boers. Without waiting to find out and paying no attention to an officer with a white flag they started to attack. When the fight looked like getting serious a few shells were put over their heads by the section of guns with the rearguard, which brought the Yeomanry to their senses. They bought their experience with three men slightly wounded.

On 15 May, on reaching the junction of the Klip River and the Vaal, the column moved south along the valley of the Klip, De Lisle being back again in command of his own column. The country was rugged and the Boers constantly attacked the flanks and rearguards. Night raids by a squadron or MI company to a likely drift or to some outlying farm took place daily. On one such raid the cavalry major, in charge of a squadron of dragoons in a neighbouring column, successfully surprised a small party of Boers asleep in a farm, including the commandant of a local commando. Whilst the other four or five prisoners were easily secured, the commandant hung vigorously to the bedrail at the foot. Instead of skinning his knuckles with the heel of a boot or a rifle butt, which would have forced him to let go, the major and he had a glorious tug of war in which the major was worsted. Other Boers in the neighbouring farms of the hamlet woke up and the major was compelled to come away empty handed, to the great amusement of all Elliott's force, as the story rapidly spread. While the columns worked in the rugged glens of the Drakensberg, Colville's Imperial Yeomanry drove the great herds of captured stock along the narrow Klip Valley which runs between the great range and its outlying range of the Gemsbokburg. The Boers holding the passes of the Gemsbokburg in particular put up considerable resistance. Lowe, who had taken over from Bethune whilst the latter was acting as Chief of the Staff to Elliot, had to call on De Lisle for assistance before he could force his way through. On 19 May the stock captured by De Lisle: some 3,000 horses, 700 cattle and 16,000 sheep, were driven over the Klip River by an improvised bridge made of waggons and sent with two companies of the 6th as escort to Bothas Pass, 'D' Company being further entrusted with twenty-three Boer prisoners destined for Natal. On arrival at the pass the stock was duly handed over to Colville to join his great herds. Not so, however, the prisoners, as a quiet unauthorised opportunity to slip down to Ingogo Station, on the Natal railway, and to personally hand them over there presented itself. Bothas Pass from the Free State side appears merely as a gap between two low grass hills on a steeply sloping ridge of downs. Beyond these two little hills, however, the range falls in the steep escarpment of mountain to the plains of Natal, the road winding down a mountain spur to Ingogo, just visible in the far distance. On the two small mounds were the various column commanders with Elliot and his staff making reports of the drive. The company with their prisoners filed by as quietly and unobtrusively as they could. General Bethune, however, spotted them and sent a staff officer to fetch the OC company to enquire as to his mission. Arrived in front of Bethune the officer was beginning an apologetic reply about having prisoners to hand over. Bethune, upon seeing who it was, laughed and said, 'So you want to go into Natal – got the Gordon Company under you?'

He replied, 'Yes sir.'

Bethune prompted, 'Only the Gordons?'

Again, came the reply, 'Yes sir.'

'Alright young man, slip along quickly out of sight before General Elliot sees you,' and away the company plunged out of sight and mind, not to reach

Ingogo till nearly dusk. There receipts were taken, with the station written large by special request in order that the company might qualify for the Natal bar. Camp was not regained till midnight, but the company was used to lone treks in the dark and Boers were unlikely to be met in so general a concentration of troops. The company later found they had qualified for the Natal clasp, the only troops in De Lisle's column to do so and with Bethune, an ex-Gordon, to thank for the opportunity. The explanation of the late arrival back was not enthusiastically received by the column commander, who thought the company had got lost.

Before entering Harrismith to refit, Elliot next directed his attention to the *Wittekopjes*, an amphitheatre of hills at the southern extremity of the Gemsbokburg and some 25 miles from Botha Pass. This strong natural fortress was not, at the time, marked on any map and had somehow so far escaped attention. Some 400 Boers were reported to be in hiding there with a large amount of stock. On near approach it was found to resemble a gigantic Norman castle, the precipitous face of the hills taking the form of giant bastions and a keep. Like so many of these hill strongholds it had few roads of ingress and egress. Admirable for a siege defence, it was liable to prove a trap for converging columns. Memories of Prinsloo's surrender overshadowed the whole district and finding Elliot's columns converging on them from three sides, the Boers slipped away as discreetly as they could. 1,000 good horses had nevertheless unavoidably to be left in British hands. That night was bitterly cold when Captain Shaw of 'C' Company, going to his hut before dinner, found a cobra coiled up in his valise. Removing himself quietly before the snake was disturbed, it was shot with a gun and then thrown writhing on the blazing mess fire, to make certain it was dead. In a supreme effort the snake nearly got one of the group of officers standing around before it fell back dead.

Harrismith was entered in triumph with 1,700 horses for the Remount depot and a two-day halt was made for supplies. Some 190,000 head of stock had been sent into Natal as a result of the raids, in addition to the 300 vehicles captured and the refugee families already there. Only the aforementioned twenty-three prisoners had, however, been captured by these columns combined. Leaving Harrismith on 25 May, Elliot started back on the return journey to Kroonstadt, clearing the country of supplies en route. The scope of the return raid was to be the country east of the Wilge River as far north as Vrede and thence via Reitz and Lindley to Kroonstadt. Large herds of sheep were encountered on the way. Already driven hard by the Boers it was found impossible to bring these poor animals along even at the slow rate of progress of the columns clearing the country. They had to be slaughtered at the midday halt and at camp. The work was generally done by natives with knobkerries but, at times, the main body of a company on detached duty or on rearguard had to do the unpleasant duty with the bayonet during a halt. It was a pitiful business. Naturally choice bits of mutton were to be had for the taking. Kidneys and chops were a glut on the market. Resistance was met the first day out and several sharp encounters took place, assistance on two occasions having to be asked from another column, with the different *spruits* athwart the road to Vrede being held in turn. Several attempts were also made to cut off the rearguard, undercover of *veld* fires. An especially stiff scrap took place near Cornelius spruit where the Boers had to abandon a big drove of 28,000 sheep which the column found themselves compelled to kill that evening in camp.

Vrede was reached at dawn on 29 May, after a very cold night march, and found empty, but the baggage guard, arriving later, was fired on most of the way. From Vrede the column turned southeast halting on the banks of the Wilge on 5 June and capturing a small *laager* of twelve waggons with a good head of cattle in the fog before arrival in camp. The natives of the *laager* revealed the news that a very much larger convoy was moving south, escaping from the more northerly columns and some 15 miles to the west of the river. Sladen, with 200 of the 6th MI consisting mainly of 'A' and 'D' Companies and 100 South Australians attached, was dispatched in the early hours of the morning to intercept them. The adventure, before the day was out, was to prove one of the most costly actions in which 6th MI were engaged. The convoy was duly sighted shortly after daylight and the usual wild gallop by two lines of horsemen down the waggons shouting 'hands up' followed. Surprised and outnumbered the Boers made no resistance and the convoy was parked near some *kraals* on Graspan Farm. Most of the occupants of the 120 waggons were women, among them a Red Cross nurse from Holland who complained bitterly and was rather curtly told to 'shut up'. Of the male combatants some managed to escape at the first rush but forty-five armed burghers were captured. By the time the *laager* had been duly parked, the women pacified, it was nearly 10.00 pm and Sladen sent off sixty of the Australians to inform De Lisle, whose advanced guard was due to arrive an hour later. Meanwhile a small party of escaped burghers had ridden off to Reitz some 7 miles distant where, unknown to the British, de la Rey with a picked escort had come to confer with Steyn and de Wet. The two latter were escorted by Commandant Davel, leader of de Wet's bodyguard. The messenger found the Boer *laager*s finishing breakfast and ready to move should necessity arise. Covered by Davel and his scouts this picked body of Boer fighting men moved off to the scene of action, picking up a small commando en route which swelled their numbers, to some 200 men.

Meanwhile, Sladen's Australians had lost their way, and De Lisle hearing nothing further was fully occupied getting his column over the drift over the Wilge, which proved a very bad one. Shortly before noon, by which time De Lisle was expected, large bodies of horsemen riding in regular formation were seen on the skyline to the west and were mistaken for Bethune's column. Captain Craufurd and Lieutenant White were sent out in two directions to direct them to the captured convoy, no thought of Boers at the time being entertained. Sladen, after posting his outposts, had placed 'D' Company in charge of the prisoners in the *kraals*; the remainder of his force being grouped around the waggons. When, too late, Craufurd found himself challenged by a line of Boer scouts 20 paces ahead, lying in the grass, flight was impossible, so he threw his bandolier high in the air before surrendering. This slight warning proved sufficient for an immediate alarm and stand to. Finding further concealment impossible the Boers, under Davel, now charged down on the *laager* driving the outposts back on the main body at once, the British retiring through the waggons to the group of *kraals* and a cattle enclosure nearby. Owing to the presence of the women and children in the waggons, the troops refrained from firing through them thus enabling their assailants to come to close quarters at small loss and establish themselves close to the *kraals*. A captured sentry over the waggons was sent in with a white flag demanding surrender as the British were outnumbered and practically surrounded. As he approached the *kraal*, Sergeant Rothnic of the Gordons shouted to him to drop his flag and lie down, which he did, remaining miraculously unhurt under close-range fire during the remainder of the fight. The women and children were now got away by the Boers, along the rear line of waggons to everyone's relief. A fierce fight now raged for well over three hours at only a few yards' range. The MI had the best cover as the Boers held only a low mud wall and an outlying hut with the shelter of the row of waggons. They were, on the other hand, some of the pick of the Boer Army and all crack shots. Both sides suffered heavily.

At three o'clock De Lisle, who was moving leisurely to the convoy which he imagined safely captured and *laagered*, was met by Lieutenant White who after being stripped and relieved of his boots, had been let go. White, limping and breathless, told of the struggle and De Lisle came up at the gallop to the rescue. The fight was over and in a short pursuit most of the waggons that managed to get clear were recaptured. The little *kraal* was found to be a perfect shambles. Thanks to the gallantry and resolution of Sergeant Sutherland, Gordon Highlanders, in charge of the prisoners, only three of forty-five managed to break away. On the British side Lieutenant A. Cameron, Gordons; Lieutenant G. Strong, Bedfords; and Lieutenant T. Mair, Australians were killed and Captain Finlay, Bedfords, seriously wounded (to be afterwards invalided out of the service). Twenty-six other ranks killed, of whom ten were Gordons, and twenty-five wounded. The Boer losses were also heavy: seventeen killed and twenty wounded left in British hands. Sladen subsequently received a brevet and White a D.S.O.

Captain Craufurd had, after capture, been brought before de la Rey and de Wet. Perhaps unwisely he had answered all questions in Taal. At the insistence of the Dutch Red Cross nurse he was accused of being a Cape renegade and of having violated the Red Cross. His answers to questions being unsatisfactory and his bearing defiant, he was to be shot out of hand. Just as he was being led away, the women of the *laager* who had been silent witnesses of the cross-examination suddenly made a rush and surrounded him declaring he had done his best for them in the early morning and that he should only be shot over their bodies. It was a dramatic moment which will live long in memory, and de la Rey handed him over to two guards for further investigation. He escaped later under pompom fire of De Lisle's relief, when perhaps the erratic shooting of his two guards, who emptied their magazines at the running figure at only a few feet, was better aimed than might appear. On the journey back to Lindley he made the comfort of the now recaptured womenfolk his special charge. The dead were buried at Reitz in the town cemetery next day. Boer prisoners, women of the *laager* and troops alike attended the service, which was held in common for the fallen of both sides.

From Lindley a sudden break back was made towards Bethlehem in the hopes of capturing a *laager.* 50 miles were covered in the next twenty-four hours and two small *laager*s picked up en route but not the one for which the effort had been made. On 12 June, the force moved definitely into the line and Kroonstadt was reached on 16 June, flanks and rearguard being harassed all the way. The total results of the seven weeks' outing were thirty-six Boers killed and buried, 100 prisoners, 131,500 heads of cattle, 260 waggons and an immense quantity of foodstuff. In addition, some 2,000 ponies and horses had been sent to the Remount depots and roughly 100,000 sheep had been slaughtered on the *veld.* Lord Kitchener came to inspect the columns and in the light of experience gained, a lighter scale of transport was initiated with bivouacs to replace tents.

Whilst at Kroonstadt a draft from the 1st Battalion consisting of Second Lieutenant A.L. Davis and A. Craufurd with thirty-five men had joined up. Lieutenant Mitford had been sent back to the line from Vrede in the middle of May with a slight wound in the forearm, the result of playing with a Mauser automatic pistol, which he insisted on carrying, and with which he both killed his horse and wounded himself whilst firing off the magazine. He did not return to the MI. After Graspan Captain Craufurd had been appointed adjutant of the 6th MI in place of Captain Finlay, wounded and invalided, in addition to his duties in command of 'D' Company.

After a full week's rest in which polo and such other recreations as Kroonstadt offered occupied as much thought as the actual refilling, Elliot's columns started out for another month's raiding east of the line. The new raid was to take the form of a great circular sweep, first southwards by way of Lindley, Senekel and Bethlehem to Harrismith; thence north by the valley of the Wilge as far as Frankfort, then back to the line via Heilbron. The raid included a good deal of night work and, in spite of the cold, it was often difficult to keep awake. Senekal was raided at dawn on a foggy morning on 25 June but the Boers had left at midnight.

On one occasion, when 'D' Company was out raiding a couple of farms in the early hours of the morning before dawn, on the roll being called at daylight, 'Private Simpson, him wi' the big feet,' was reported missing, though his horse was present. He did not reappear till noon of the following day, having fallen asleep at the second farm while standing dismounted and, as his horse continued in the ranks, no one noticed his absence till daylight. He apparently slept the next twenty-four hours, awaking at daylight next day, and was directed back to camp by the Boer woman of the farm after a cup of Boer coffee! On another night march, when moving after a party of Boers said to be *laagered* at Kaffir Kop, a conspicuous hill and well-known rendezvous on the Lindley/Bethlehem road, the column got split into two at a small drift in a rain storm. On emerging from the drift two of the main body companies, as well as the baggage and rearguard, found themselves completely lost and, after an hour wandering, had to park till daylight and did not rejoin until 9.00 am. The irony of the situation was that a Boer convoy had got similarly lost that night and so unwittingly had escaped capture. A lot of stock was captured and thousands of sheep killed. Official diaries contained constant laconic notes: 'Killed 3,000 sheep at midday halt, 4,000 at night.'

On 2 July captured stock was handed in at Harrismith and ten days' supplies taken for the move north. Next morning Major Sladen was met by his groom with his horse and the remark, 'Here's your horse, and I'm fed up.'

'What do you mean?' asked Sladen.

'I've been and shot your servant,' was the startling reply. It appeared that, quite accidentally when cleaning his rifle, he had shot dead the servant who was his special pal, both men having come from the East Yorkshire Regiment with Sladen on appointment to the MI.

As on previous occasions the march up the Wilge valley was strenuously opposed, skirmishing with the troops protecting the convoy continuing throughout each day's march. On 6 July De Lisle as left flank column entered Reitz shortly after dawn. The Boers, as usual, had evacuated in anticipation of a British move. While billeted that night in the deserted town, De Lisle was lucky enough to pick up a document, which very nearly had momentous consequences, for it proved that the Boers were in the habit of leaving the place as the British column approached to return again as soon as the danger was safely passed. Elliot turned this piece of information to good account. Four days later when his whole line was well to the north, he directed Broadwood, who was right flank column, to double back by a forced night march of 24 miles and take the town by surprise. Broadwood moved light with 400 men and a pompom. A loss of touch by a connecting file in the dark delayed the march a little and Broadwood found dawn breaking when he was still 2 miles from the town on the morning of 11 July. He had to gallop hard to reach the town before day broke and had no time to block the exits on the far side. The surprise was complete. With the exception of de Wet, practically the whole Free State government were in the town, and all unsuspecting, still asleep. Steyn in his nightclothes mounted bareback and, with only a rope halter for a bridal, managed to escape with seven of his bodyguard. Steyn himself was hotly pursued by Lieutenant Shaw of the 7th Dragoon Guards and a sergeant of the same regiment. They were, of course, unaware of the identity of their fugitive. Shaw's revolver misfired whilst covering Steyn at 10 paces. Steyn's pony was fresh and he gradually drew clear. Broadwood's captures including twenty-nine prisoners including the whole government staff, Davel (with many of the personal bodyguard); Generals A.P. Cronje and J.B. Wessels and Field Cornet Steyn, the president's brother. £1,000 in gold and £11,000 in Free State treasury notes were found in Steyn's quarters. Broadwood then rejoined the column.

The whole force now swung westward by way of Frankfort and Heilbron at which latter place the columns concentrated and a divisional church parade for the combined columns was held on Sunday, 14 July. Kopjes Station on the line was reached on 17 July and the results of the month's raid were seventeen Boers killed and wounded, sixty-one prisoners, 7,000 horses, 7,000 cattle and 300 vehicles. British casualties, in spite of constant scrapping, totalled only three wounded.

The first true drive occurred towards the end of July when a more ambitious scheme was set in motion, worked by General Headquarters staff under Lord Kitchener's personal direction. Seventeen columns were to take part, with a total strength of 15,000 mounted men. The scope of the operation can best be likened to a well-organised beat for tiger in an Indian jungle, staged on a heroic scale. The line of the Modder River had recently been fortified by a series of constabulary posts providing a convoy line of communication between Kimberley and Bloemfontein Railways. The South African Constabulary Posts, strengthened by several mobile columns, was then to form the Line of Machans[16]

[16] The term "Machan" describes a hide used for hunting. Term often used in context with Tiger shoots.

The two railway lines similarly strengthened formed the 'stops' whilst Elliot's column represented the beaters, accompanied by a proportion of guns. The beat was roughly 100 miles wide from Kimberley to Kroonstadt and 120 miles from Modder to the Vaal at Klerksdorp, which was to form the point of departure of the beat. Those who had taken part in such work know how often an old and weary tiger manages to break through. The quarry inside was every whit as brave and resourceful as any tiger and animated by much the same point of view - safety first but fierce and determined attack when necessity demanded.

Elliot first moved to Klerksdorp for position of manoeuvre. The move west from Kopjes on 18 July was over familiar ground and the country was swept in the usual manner without much result. 'D' Company, 6th MI, however, managed to surprise and capture a small *laager* with five Boers and seventeen horses camped at a little backwater of the Rhenoster Spruit, a tributary of the Rhenoster River. Noticing a smart little pony rugged up, which was unusual, among the captures, enquiries were made and the previous owner, brother of the commandant, said it was a half-bred Basuto, sired by one of the government stud farms' English stallions. He evidently hoped the prize would fall to an officer. Close inspection showed it to be a beautifully shaped little animal and just the stamp to make a first-class polo pony. The captain earmarked it for himself if it could be got past De Lisle, which was doubtful though the rough, unclipped coat gave hope. Before moving off after breakfast the pony was bespattered with mud and a bribe of a tin of jam and some packets of cigarettes to his old master to hold his tongue. The prisoners were interested and amused. A heavyweight sergeant was put on his back and he was given a good bucketing on his way home. He came in anything but fresh. On arrival the matter was talked over. The first eight horses were to go to the MI and it was arranged that he should not be picked out till 'D' Company's second turn ie fifth choice. If he could be kept out of view till then, no special notice would be taken as it is certain that all the ponies would have been carefully inspected whilst in the MI charge. The column was not rejoined till the midday halt and distribution was arranged for in the evening. The plan worked well and the tired little pony was passed with no special notice. Once safely in the captain's possession he was clipped, groomed and thoroughly rested. Some days later De Lisle passing by on the march asked the captain where he had picked up his smart new mount. 'At Rhenoster Spruit,' answered the captain.

'Strange,' said De Lisle, 'I never noticed him or you would not have got him.'

'No sir,' came the answer, 'We did all we could to avoid notice for that very reason,' and the story was told. De Lisle laughed and added that as the MI had grown so smart he would know where to come for the next tiresome job.

The following night another small *laager* was captured by two squadrons of South Australians and two companies of MI, this time under De Lisle himself. This consisted of twenty vehicles together with eight prisoners and twenty-four rifles. In galloping the convoy at daylight, Squadron Sergeant Major McGillivray of the Australians was shot dead by a Boer woman with a rifle shooting from a Cape cart.

The drifts over the Vaal from the railway to Klerksdorp were held by infantry detachments. Coal Drift, where the column crossed, was not improved by the carcasses of 9,000 sheep slaughtered by some column returning from a foray a while back. Two days' halt was made at Klerksdorp during which some twenty men of the 6th MI persuaded the Yeomanry Regiment quartered there to exchange some of their English horses, recently out from home, for their own less restive mounts. Such an addition could not fail to draw notice and on moving out the adjutant had made discrete enquiries. This was fortunate, as he was able to coach his CO. A little while later De Lisle himself spotted the improvement in the ranks. They were, however, able to satisfy him that no actual theft had taken place as the Yeomanry officers, apparently either indifferent or unobservant, had made no complaint. After all, the 6th MI were likely to make better use of the exchange.

The drive started on 29 July, De Lisle being right flank of Elliot's driving columns to which Rawlinson had been added on the left flank and Garrat in the rear to deal with any commandos breaking back. Whilst along the Vaal, that is to say as far south as Hoopstadt, night raids to the drifts were the order of the day and several catches were made. On 30 July, Shea with 200 South Australians captured a *laager* on the Valsch taking twenty prisoners, twelve vehicles and stock. 'B' and 'C' Companies on another drift on the same night having no luck. Next day, while 'D' Company went to the junction of the Valsch and the Vaal on a fruitless mission, Lowe captured two small *laager*s with eleven prisoners and ten waggons higher up the river. The most important raid was that of Shea with 200 Australians on the Vet River on 2 August. Starting at 11.00 pm the night before, having been given information, he reached his *laager*, all unsuspecting at dawn, at Groot Vallei near Hoopstadt. Galloping the *laager*, he had the bad luck to run into a wire fence thus giving the alarm and the Australians had to charge on foot. The *laager* proved to be that of General Smuts and 340 Transvaalers, then on their way to join Kritzinger in Cape Colony. Owing to the warning given by the wire, Smuts and the majority of Boers managed to get clear. Six Boers were, however, killed in the melee, twelve captured and a lot of horses taken. After daylight, seeing the size of the attacking party, the Boers rallied, but as the counterattack began to develop, the whole of the 6th MI, with a pompom under Fanshawe, arrived in view and the Boers cleared off. The MI had started at 4.00 am to reinforce if necessary. Two Australians were killed and one each of the Australians and MI wounded in this little fight. Broadwood captured sixty prisoners and 4,000 head of cattle the same evening.

Unlike the Eastern Free State, the flat open nature of the country allowed for a very loose formation and the great frontage covered by the driving columns necessitated very wide extensions. The flanking companies of the column often worked 3 or 4 miles away from the main column itself. At night, however, columns were necessarily camped 10 or 12 miles apart. Although night raiding parties to the drifts of the various rivers in the line of advance, and to any other likely camp sites, did make movement difficult, many small parties of Boers did, undoubtedly, break through. During the daytime the Boers had to be continually on the move, as any standing *laager* was only too likely to be spotted and fall prey to the British. It was a country of occasional mirage, and moving to the Pan, 30 miles west of Boshof, the column was misled several miles out of its way marching to the imagined water.

On 8 August, two companies of the 6th MI and two squadrons of the South Australians left at 2.00 am and after dawn had a great chase after fleeing Boers, capturing over 100 waggons, 400 head of cattle and prisoners.

On 10 August, the Modder was reached, and the river was crossed at Abraham's *kraals*. From here the column swung left along the river to Glen Lyon Siding, which was reached on 13 August with a five-day halt and an inspection by General Alderson, Inspector General of mounted infantry. The captured stock was sent into Bloemfontein and a certain number of passes were issued into the town. The total results of the drive had resulted in seventeen

Boers known to be killed or wounded and 259 unwounded prisoners, besides 814 waggons and 21,000 head of cattle driven in, and 186,000 sheep of which a large number had to be killed on the *veld*. There was some soreness about the cattle as these were handed into the commissariat and sold to the cold storage company at nominal prices, to be later issued to the troops. Originally a legitimate enough enterprise for purchasing livestock in the Cape Colony and delivering to the troops in the field, the contract should have been terminated once the policy of denuding the country of stock was initiated. Its continued existence was one of the scandals of the campaign. Smuts himself, after his encounter with Shea, had managed to break back and crossed into Cape Colony a fortnight later after that drive was over.

After the short rest at Glen, Elliot moved east to the Basuto border. For the first three weeks the move followed roughly that of Hamilton's march of the previous year. Barker and Pine Coffin were co-operating from Winburg.

On 22 August, a march was made on the Korannaberg, soon to be a familiar haunt. Among the many natural hill forts of South Africa this big, isolated mountain is perhaps the most perfect example. On it in previous years the Koranna tribe had held out successfully for seven years against the Boer commandos till they were eventually given safe conduct to Basutoland.

Some 10 miles long and 5 miles across it is shaped like a gigantic kidney. It is entirely flanked by cliffs except at the pinch of the kidney, which faces Basutoland to the east. Here a good cart road runs up to the plateau above but this in turn is flanked by cliffs. The northern half of the kidney consists of a plateau of good-looking land overlooking the Free State plains far and wide. The southern half consists of an inner valley entirely shut in by the hills of the outer rim. There is just room for the road to pass with comfort into inner valley, which is drained by a stream carrying off the water over a series of falls into the plain below just where the road first starts. In the inner valley the Boer owner had his farm, entirely shut out from the outer wall. On the western face there is a practicable goat track, scarcely a footpath, which leads through a split in the cliffs just wide enough to admit one man at a time onto the northern plateau. The existence of this path was known to few Boers and had not been used since Koranna occupation days.

'D' Company was sent on ahead the night before, with a Boer guide, son of a retired French general of the 1870 campaign, who knew where the path was supposed to start but had never been up it. There was a good moon and 'D' Company's orders were to try to establish themselves at the little gap by dawn, if possible, leaving their horses at the foot of the hill. Unless discovered they were not to give themselves away and if the path proved impractical, they were to come away before daylight when the main body was to be at the main road. On arrival at the foot of the hill the goat track was found to have disappeared except as a path for wild animals, but the gap above could be seen as a guide. It was a scramble over huge rocks carried out as silently as possible. When near the top, with the first flush of dawn, suddenly from down below came ringing up in a strong Scotch voice, 'Jock man, Jock are ye hurt?' Somebody had fallen on the rocks below. Horrified, the captain, one of his lieutenants, the colour sergeant and one or two others, far ahead of most of the company, made a dash for the gap then plainly visible 100yds ahead. As they did so, a leopard descending brushed quietly past them, actually touching them with his flanks on the way. The gap was gained, it was just a narrow gateway of stone, and the party breathed more freely. The Boers, however, had quietly left earlier in the night, fearing to be trapped with the British column in the neighbourhood. The led horses had, in the meantime, rejoined the column and the captain, with others, rode over the fortress with his mind, for the moment, back in India dreaming of Gwalior and Chiter.

The Korannaberg is an outer bastion of the Wittebergen, with memories of Pinsloo's surrender, and these hills leading to the Brandwater Basins were next searched with General Barrington Campbell from Ficksburg co-operating.

A halt for three days was made at the little hamlet of Clocolan to get supplies, spare time being occupied by polo and an inter-brigade gymkhana, the remainder of the month being spent in clearing the Wittebergen.

One day when clearing the Wittebergen hills under Fanshawe, a cloud of locusts appeared in sight. Fanshawe mistook these for a distant commando of Boers and, calling in the outlying companies, he got his pompoms in position for a fight. The arrival of the locusts shortly afterwards caused great joy and it was a long time before he heard the last of his commando.

At the end of the great drive in the Western Free State, a new ordinance had been published, stopping the burning of farms, except as an isolated preventative measure, and in consequence the women were again able to live in them, providing they could get food. There was an immense amount of grain and flour untouched, and the Boers managed, between raids, to plough and sow large areas of the country, especially on the borders of the Drakensberg, where the movement of columns was not so frequent as on the plains around railways.

A good deal of time in the operations of the next month or two was spent cutting standing crops of wheat, maize and oats, the latter being fed to the horses as oat hay. In camp, an outlying picquet would often be sent to hold commanding ground within a mile's radius. Where there was a farm, the women would often volunteer to cook the officers an evening meal. Liberal rations of tea, flour and sugar were handed over and a convention sprang up that a new tin of jam, butter or condensed milk should be opened, of which the officers used only a spoonful or two. Where only one smart dress remained, the younger women would take it in turns to don their finery and entertain the officers. Most of the younger women had had an English education at school, being more fluent than their brothers.

From Wittebergen, Elliott's column moved into Winburg to hand over captures and get fresh supplies. The district had already yielded 100,000 head of stock, 140 carts but only nine prisoners, and the raid was to be repeated. During the five days' halt a combined gymkhana was held by Elliott's three columns.

The usual night operations were carried out by the 6th MI on leaving Winburg but without much success. However, the second day out on 11 September, De Lisle, with two companies from the 6th MI and one squadron of South Australian, had a fine galloping chase after a Boer *laager* giving a good 12-mile point and ending in the capture of the convoy with seventeen prisoners, sixty-nine carts and a lot of stock. If hounds were missing, the ride over the open space in full pursuit of the vanishing quarry was not a bad substitute for a good hunting run.

The column was now back on the Basuto border and night raids were frequent. After a weary ride through most of the night to Nels Poort Pass, through the Wittebergen, the 6th MI had to rejoin their column on the move and with some diffidence, as they had completely lost their way.

There were many big native locations near the Basuto border and it was part of the government policy, when clearing suitable horses, to give native owners a requisition form up to £15 or £20 for a really good animal, to be later exchanged by the owner for cash at one of the garrisoned towns. On one occasion among a small batch of suitable ponies, a headman produced one, almost own brother to Frickie, the Basuto pony. Captain Craufurd was acting as interpreter to the column as the official interpreter to the column was absent. Suspecting so good a pony could not really be the property of the man, but merely placed for safety in his charge by the Boer master, he asked the Remount officer if he might offer two sovereigns, which he happened to have on his person, in lieu of the government requisition order. These were probably two of the very few in the camp as gold was seldom carried. The Remount officer laughingly agreed, providing he explained the value and the meaning of the government note, and never suspected the headman would take the gold and regard the offer more as a joke. To his surprise the offer was accepted with alacrity and as the transaction was quite voluntary the purchase was duly certified later by the OC column. Truth to tell, the two sovereigns could be hidden in a loin cloth with no one the wiser, the real owner being merely told the pony had been commandeered. The requisition order could not have been cashed without the fact becoming common property. A week or so later the captain rode into Harrismith with his two ponies to get the purchase officially ratified. He asked the OC Remounts if, for sentimental reasons, the certificate could be transferred to Frickie, whose mettle had been fully proved at the Winburg gymkhana. 'Certainly,' said the OC Remounts, '… as far as my department is concerned, but why not buy in the other pony for £15 if you are entitled to that number of private chargers.' The cheque was drawn forthwith and the captain rode back with certificates of ownership for both ponies. Privately owned remounts entitled the officer owner in a mounted unit to a small horse allowance. Both ponies turned out very good polo ponies. Frickie was later sold after peace was agreed, to a cavalry officer of the permanent South African garrison for £80.

After the first week, Elliot's columns were diverted to the Natal border owing to Louis Botha's threat to the Natal. On the road into Bethlehem, De Lisle's column picked up one wounded man of the 7th Dragoons, buried another and bought in four released prisoners, the casualties of a small scrap Broadwood had been in the day before. The next day, 22 September, Broadwood was again heavily engaged while De Lisle's column marched into Bethlehem. The garrison of this little town lived a life of semi siege, often being cut off for weeks at a time from all communication with the outer world except by helio.

On arrival at Harrismith on 26 September, orders were received to train for Zululand where the Zulus had cut up a Boer *laager* which had incautiously entered their territory. In accordance with long tradition every living thing had been ruthlessly slaughtered including men, horses, oxen and fowl, and the Zulus were threatening to take the field en masse. The political officers were, however, successful in stemming the tide and orders to entrain were countermanded, Elliott being diverted to the Natal border, along the Drakensberg, instead.

Elliott's booty on entering Harrismith consisted of thirty-six prisoners, 60,000 head of cattle and 228 waggons. 400 worn-out Boer horses had, in addition, been destroyed the day before entering the town. The column got 100 remounts before marching out again.

For the next fortnight the columns were to operate in the country enclosed by the Cornelis and Molen River as well as in the upper reaches of the Klip River. It was wild rugged country with numerous fertile farms in the valleys and had been little touched by operating columns. For six months it had been unvisited. Crops were ripening everywhere, and the country was full of stock and Boers. Every day parties of two squadrons or more were out on their own, after leaving at night and returning on the afternoon the following day. This gave many an opportunity for scraps with parties of Boers from 100 to 200 strong, but there were too many parties out, and the combined columns were too strong for the Boers to put up any very determined resistance. The African summer had commenced, the fruit season was in full swing in the fine orchards of the farms and the destruction of crops was rather in the nature of a picnic, with the excitement of the day's scrap thrown in.

On 3 October De Lisle took 300 men from the 6th MI and 100 South Australians, two guns and two pompoms to Mullar's Pass to get in touch with Newcastle, leaving a standing camp behind on the lower Molen River and travelling light. The party was away three days. The second day out De Lisle selected a cup in the hills, not unsimilar to the Bull Hoek in Cape Colony, as camp. All three companies of the MI volunteered for picquet but the idea was turned down – however, all slept fully armed and accoutred. Shortly after midnight De Lisle came to the adjutant's bivouac to tell him to reinforce the picquets an hour before dawn – the adjutant, who was asleep, woke with the voice and the next thing De Lisle knew was the adjutant's rifle on his stomach, finger on trigger and a gruff challenge as to who he was. Considerably startled, De Lisle said next time he wished to wake the adjutant in the night he would send an orderly!

On 12 October, moving north through the Wittekopjes to meet a convoy with supplies, the column ran across Louis Botha and a bodyguard of forty-five men returning from Natal. The Boers were on the north bank of the Hol Spruit and, in forcing the passage over the drift, the Australians had two men wounded and one captured whilst the Boers lost two killed and two wounded but Louis Botha himself got clear. Three days later after getting supplies, a small round-up was tried on the banks of the Wilge. Three separate squadrons were sent to various likely *kopjes* during the night while De Lisle with two squadrons, two companies and two guns moved off at 9.00 pm. Around 14 miles were covered by 2.00 am and a further march was made shortly before dawn. The move was successful, for after daylight, fifteen prisoners and thirty vehicles with a lot of good ponies and cattle were rounded up without a fight. Camp was regained at 3.00 pm. Another similar move was tried the next day. De Lisle with the main force drove the Boers,, after some fighting, on to the columns; Fanshawe hearing the guns, turned out the column troops to meet them. The fight looked like proving a regular trap. Reinforcements of Boers, however, came up from the fastnesses of the Wittekopjes and a stiff running fight ensued. The MI were outnumbered and eventually had to retire. In the rearguard fighting Lieutenant Heath of 'B' Company (Welsh) was wounded. Four Boers were killed.

On 19 October, just as the usual night raid of two squadrons and two companies were parading to start off at 9.30 pm, heavy sniping broke out all around camp and the camp stood to arms for an attack. Nothing further matured, however, and, after an hour's suspense, the party moved off. A further delay occurred owing to a heavy rain storm during which the column got split into two at a drift and another two hours were lost. Consequently, on arrival at 3.00 am the *laager* had moved and there was nothing for it but to bivouac till daylight to eventually get back to camp at 4.00 pm.

For the next week there was constant night raids and daylight scraps in the district south of Vrede with one or two minor casualties every other day on both sides, including an officer of the South Australians. From Vrede, Elliott's force moved up to Standerton for a short rest arriving there on 27 October.

Since leaving Harrismith the combined columns had secured thirty-three Boer casualties in killed, wounded and prisoners accounted for, eighty-three vehicles and 26,000 head of stock. Standerton was full of columns returning from Botha's raid into Natal.

Whilst sitting quietly at tea the day after arrival, the officers suddenly heard a shout of, 'Turn out the 6th MI, the Gordons are fighting the Australians.' Obedient to the summons the officers too turned out and found a glorious fight in progress. The Australians had purchased two barrels of beer which the MI had suggested they should share. The Australians demurred and, in the argument, one of the barrels had got broached. A scramble and fight for the beer commenced during which Private Fraser, a very powerful man and now a highly respectable citizen of Aberdeen, broke one of the Australian's arms; a pause ensued while both sides called for reinforcements. It was at this stage that the officers arrived and managed to restore order. A present from the Officer's Mess of a new barrel to the aggrieved Australians fortunately resulted in a combined sing song before higher authority heard anything of the fracas.

The rest at Standerton was destined to have a rude ending. Suddenly, at noon on 3 October, news of Benson's disaster at Bakenlaagte, 50 miles to the north, the day previously came through by helio and all fit men and horses of De Lisle, Hamilton and Allenby's columns were ordered to parade at short notice. Five days' rations only were to be carried and no other baggage. This meant only two mule waggons per column. Two hours later the force was on the move but owing to the sketchy orders there was considerable confusion and delay at the bridge over the Vaal. A halt was made for three hours just after dark at a hamlet called New Denmark to enable the now loaded supply waggons to catch up. At 10.30 pm the force moved on, the night being very dark and gradually changing to fog. An hour's halt was made at 4.30 am to feed the horses. The fog, which was now thick, did not clear till noon. On moving off in daylight the force shook out into fighting formation with De Lisle on the right, Allenby centre and Hamilton left. Quite early in the morning fighting started in the mist, first on the left and then right along the line to the left and quite a lot of Boers came into view.

Helio communication was established with Benson's column first at 8.00 am, during breaks in the mists. Wools-Sampson, who was in command after Benson's death, was able to report the column was holding its own and entrenched. A halt of two hours for reconnaissance was now made near Kromdraai where there is a break in the low line of hills intervening between the beleaguered column and the relieving force. Not much resistance was offered as the Boers, under Botha, having failed to storm the camp, were only too anxious to get clear away with the captured guns before becoming too heavily involved with the different relieving columns. When Benson's camp was finally reached at 3.30 pm only a scattered rearguard of Boers hung around who speedily made off.

Comment has been made that there was no pursuit, but 55 miles, as the crow flies, had been covered during the 24 hours, to which another 10 miles may be added for a skirmishing etc. The Boer horses had practically had two days' rest and only Sampson's severely handled column was available with 24 hours' hard fighting behind them. The relieving force off saddled and grazed till dusk, then moved into Sampson's outpost for the night. The march back took two days and was much delayed by heavy rainstorms and the breakdown of the two waggons. They consequently went into bivouac five miles north of Standerton at 11.30 pm marching at 4.00 am to rejoin the standing camp at 6.00 am on 4 October.

On arrival Major Sladen heard he been promoted to another billet and Major Bradley Dyne of the Buffs took over command. Major Dyne, though personally popular, was more in need of a rest at base camp than physically fit for command of a mounted infantry battalion in the field and he only held command for a few weeks.

A concentrated net of driving columns was now tried in the North Eastern Free State somewhat similar to that in the Western Free State in July; fourteen columns in all were to take part including the 9th and 15th MI. The scope of the net was a parallelogram with its four corners at Winburg, Heilbron, Standerton and Harrismith. The diagonal 175 miles long and the sides 100 miles across.

Starting on 6 November the drive was only to last six days and the intention of the move was kept a profound secret to all but the column commanders and their staffs. Two days out from Standerton 'D' Company, on rearguard, was heavily attacked and the point under Lieutenant A. Craufurd nearly rushed. As the column was moving into camp a confused group of Boers and Britons were seen riding helter-skelter over a ridge too intermixed to shoot. The Boers, on recovering from their surprise at nearly hitting the main column, hastily withdrew. One of the prisoners when handing over his carbine, a sporting Lee Metford, said it had been captured by him in the Jameson raid of four years previously. The trophy was handed to the company commander who carried it in place of his army carbine for the rest of the campaign. The part assigned to Fanshawe, temporarily commanding the 6th MI, South Australians and a company each of the 7th Dragoons and King's Dragoons, was to hold the drifts over the Wilge, at that time more or less in spate from recent rains. The different squadrons and companies were pretty constantly outlining the river. Craufurd was temporarily in command of the mounted infantry. The Wilge was crossed and re-crossed during the operations and one evening, coming into camp with two old Boer prisoners in a Cape cart, the men, fearing the Cape cart would upset with the flood, which was up to the saddle flaps of the riders, sprang out and were drowned.

Lieutenant Wilmar of the Bedfords had been wounded three days before and died of his wounds next day. Coming into camp, dead tired that night Craufurd casually reported to Fanshawe, who was in bed with fever, that he, 'Had drowned the two Boer prisoners.' Knowing there was a good deal of bitterness at this moment over Wilmer's death, Fanshawe fairly jumped out of bed, fearing an atrocity as reprisal!

Coming in late off patrol with his squadron the captain of the 7th Dragoons found himself orderly officer for the day. He was pretty tired and his cousin commanding the mounted infantry offered to do lights out for him and let him turn in. It was the brilliant moonlit night and, in one small bivouac in the cavalry lines, was a signaller reading his letters by the stump of a candle. 'Put out that light,' said the officer gruffly.

Back came the immediate reply from the bivvy, 'Yes, but first put out the blinking moon'. The amused officer passed on the reply, to the scandal of the squadron sergeant major.

In spite of the secrecy maintained and the very efficient arrangements for the drive, the Boers somehow got wind of what was on. The news flashed from hill to hill by helio and the scattered commandos dribbled slowly through the gaps at night before the cordon had got into really close touch. On 12 November when the net closed round Reitz nothing was found. In previous small skirmishes, however, thirty-four Boers had been killed or wounded, eighty-six prisoners captured and 200 waggons with 10,000 head of cattle taken. Elliott then moved into Harrismith with his captures arriving on 16 November.

Early in July, Craufurd had sent his first charger into Harrismith while camped near there, for a rest suffering from bad cuts with barbed wire. The horse was well known in Harrismith as a winner of races, having been there with Lieutenant White before Craufurd went into the mounted infantry. For the next three months, whenever the column came to Harrismith, the horse had, the day before, been sent to the grass farm 12 miles out. This time, determined to be in time, a helio message was sent two days before, followed by an officer entering a day ahead with a special party, who carried a letter to the major of the remounts. The horse was duly sent to the farm all the same. This time there was a halt for three days and the major demanded his horse. The remount major said it was impossible to get it back – a strong interview followed in the office. The remount officer finally adding that he had had the horse branded with the government arrow and that it was now a remount and not private property. The irate owner finally lost his temper and called the remount major a 'Damned horse thief' in front of his office staff. The remount officer reported the matter at once and both officers were brought up before General Rundle commanding Harrismith with Elliot and De Lisle in attendance. After hearing the evidence General Rundle turned to his officer and said, 'So Major Craufurd called you a damned horse thief, but why do you object for it seems to me that is exactly what you are!' The crestfallen remount officer retired while Craufurd remained for a private telling off. Next day, with his horse back in his own possession, the two officers dined together to laugh the incident over between them.

The mare in question had a bit of a history. The son of a clergyman well educated and equally well-intentioned, Private Dunne had proved somewhat of a problem in peace. Before CO's inspection he would work zealously at whitewashing fatigues, to be carefully hidden out of sight before the CO arrived, only, alas, to reappear at the most inopportune moment. Filled with good intentions, his captain had often tempered the scandalised wrath of unsympathetic NCOs, for whom he cherished much unexpressed gratitude. In war he found his opportunity, for he was a born freebooter. In early stages his zeal for officer's mess had nearly landed all concerned in difficulties. Then, at the formal entry of the Guards Brigade into Pretoria, a rugged Highland figure, with hose tops at heels, managed to accompany the leading battalion, determined that the Gordon Highlanders should be duly represented. His great ambition was to secure a horse for his company commander, whom he knew to be a keen rider. Here again zeal outran discretion but his chance finally came at Doornkop. In the final charge he ran on ahead and, seizing the best mount he could see, a smart little grey, shot the owner and annexed the steed. Strapping the bridal to his person he spent the night tied to his capture, which he proudly presented to his captain at the rather rugged reveille after the battle. General Smith Dorrien heard the story and interested himself at army headquarters to regularise the action. A certificate of private ownership was granted, coupled with the proviso that the pony must be exchanged at the remount depot for a sick horse, of the officer's own choosing. With a sympathetic help of the remount authorities a fine Australian mare with bad sandcrack was selected, many months later to become the first charger of the commanding officer of the 6th MI.

Private Dunne accompanied his captain to the mounted infantry, where he was in his element. A fine natural scout, he developed into a good patrol leader and was eventually promoted sergeant in army orders for gallantry by Lord Kitchener. The responsibility proved too great, and one final disaster overtook him. He was persuaded to accept a three-month bonus of his rank and revert to corporal. He ended his service in that rank, not long after the return of Regiment to Home Service.

Whilst Elliott was at Harrismith, a new move had begun on the part of the Free State Boers. De Wet had been resting on his laurels since the flight from Cape Colony in February, but Louis Botha, having returned from Natal, a letter of expostulation was sent by the Transvaal government who felt that the Transvaal leaders were now bearing the brunt of the resistance. De Wet consequently gathered a commando of 700 men to begin a new series of operations.

As soon as this information reached headquarters, Elliot was directed to deal with the situation and a fresh hunt for de Wet began. For the first week, operations centred round the Bethlehem district and two raids were made on the Langeberg, another of the cliff-encircled strongholds in which that part of the country abounds. The Langeberg is a narrow table mountain with water at the top and some 10 miles in length. Like all these strongholds, the Boers immediately evacuated as soon as there was real fear of encircling columns forming a net. In spite of a good deal of hard work not much was achieved and after a week out the force moved to Bethlehem for provisions. Whilst camped near Bethlehem Lowe's picquets, when taking up the outpost line, came into collision with De Lisle's outposts and owing to some sort of jamb the Colt gun ran out its belt before it could be stopped from firing. It very effectively scattered Lowe's picquets for the time being.

From Bethlehem Elliott moved into Kroonstadt searching for de Wet. Kroonstadt, which was now to become the centre supply, was reached on 30 November. A small convoy was caught two days out after a 30-mile night march, thirty waggons and 2,000 cattle were captured but only one prisoner. Elliot's bag for the month was eleven prisoners, 200 vehicles and 42,000 head of cattle.

Chapter 9

6th MI, blockhouse drives

Whilst Elliot was refilling at Kroonstadt, Remington had accidentally stumbled on de Wet's concentrated commando of 700 men. A drive on new lines was therefore planned which was to form a rough model for the drives in the following spring, which eventually brought the campaign to a close. Two blockhouse lines had been pushed out into the Eastern Free State to give communication with occupied towns of Lindley and Frankfort far out in the *veld*, away from railway communication. These two blockhouse lines were to form the flanks of the drive. That from Frankfort to Heilbron and prolonged along the railway to Woelverhoek Junction was to form the northern boundary. That from Lindley or rather from Kaalfontein bridge, 20 miles west of the town, which was as far as it then reached, forming the southern boundary which was prolonged eastward by Colonel Pink's small column (that included the 9th MI who were covering the extension of the blockhouse line eastward). The main railway was to form the butts and the Liebenberg *vlei* (a shallow minor lake, mostly of a seasonal or intermittent nature), the starting line for the beaters.

With a view to confusing the enemy a new feature was introduced. The columns taking part, including De Lisle, Broadwood, Byng, Damant, Remington and Wilson moving from different points on the flanks of the drive, all crossed the Liebenberg *vlei* as though they intended driving eastward. Then, on 11 December, turned and recrossing the *vlei*, drove westward. Unfortunately for the success of the plan, Broadwood came into contact with de Wet's main body at Quaggafontein on the River Valsch near Lindley on 8 December.

Finding de Wet in a strong position, Byng was called up to support and the two columns together drove de Wet and the main portion of his commando out of the scope of the drive altogether. Touch was unfortunately lost, and it was thought that de Wet had broken north. Elliot consequently concentrated on Elands Kop to the northeast with, of course, no result whatever. It was now realised that the drive had fizzled out into a fiasco and on 14 December it was called off.

News came through that it was to a kop on the Lindley Bethlehem Road that de Wet had gone and a concentration was ordered on Kaffir Kop at dawn on 17 December. From this kop, de Wet commanded a view of the whole country around and the move of the convening columns having been reported, he dispersed his commando with orders to reassemble in the Langaberg. Finding the kop empty Elliot started back through Lindley towards Kroonstad along the blockhouse line. When at Doornkloof, 30 miles out, he was however directed back to assist in covering the building of the Lindley blockhouse line, which was then pushed on with all possible speed.

Throughout December the 4th Imperial Yeomanry had been attached to De Lisle's column and their independent interior economy caused some amusement to the regular troops. One evening, long after dark, 'D' Company arrived at the new camp. The first troops struck were the Yeomanry, and the captain, riding up to a small party cooking, asked if they could tell where the mounted infantry lines lay. 'Don't know,' was the curt reply.

'Which way does the camp face then?' asked the officer. Again, they did not know. 'Where are your officers' tents?' asked the captain in despair, meaning to ask of the officers. Again, they did not know. 'But surely you know where your own officers' tents are?' objected the captain.

'No,' was the reply, 'we only look after ourselves.'

Whilst camped near Quaggafontein, covering the blockhouse line and to stem over Christmas, suddenly, on 26 December, came the startling news of the disaster to the Yeomanry at Tweefontein near Bethlehem in the early hours of Christmas Day. The 11th Yeomanry were part of a force under Rundle covering the construction of a railway blockhouse line between Harrismith and Bethlehem and held a strong bluff facing southward with its rear face sloping gently down towards the projected blockhouse line. Too cheery a view of the date had been taken and the bluff face, which should have been impregnable, was very indifferently watched. The Boers, deliberately choosing the most difficult face, rushed and captured the camp at 2.00 am on Christmas morning. Too late, a hopeless resistance was offered in the dark. Nine officers and forty-nine men were killed, six officers and seventy-eight men wounded, three officers and 203 men made prisoners with a gun, pompom and all the horses, baggage and other camp booty. The Boers lost three officers and eleven men killed, thirty wounded including one officer. The Boers moved off with their captures at 4.00 am but, before doing so, de Wet allowed the local commandos to partake of the Xmas cheer all ready for that day, whilst his bodyguard collected the Yeomanry horses. Thereby hangs a tale to be told in due course.

Elliot, for the time being, only had two small columns organised under De Lisle and Fanshawe as most of the cavalry had gone on from Doornkloof to Kroonstadt to refit. De Lisle's column consisted of the 1st, 3rd and 7th Dragoons and the 6th MI, Fanshawe's the 4th and 12th Yeomanry and South Australians. These two columns marched at dawn on 27 December. De Wet, meanwhile, moved north to Reitz and leaving Prinsloo in command of the troops had gone himself to report to Steyn. Moving by forced march through Blijdschop to the Liebanberg *vlei*, which was reached at 11.00 am on 28 December. The crossing took some time as the usual drift was out of repair and much broken. Elliot later came on Prinsloo's rearguard hurrying to cross the Rhenoster at Armstrong's Drift. At sundown he was brought to a standstill by the Boers in a strong position and reported to be 2,000 strong. This estimate doubled Prinsloo's real numbers but imposed caution, and darkness fell after some indiscriminate skirmishing. In the night Prinsloo stole away. This was the last time a Free State force 1,000 strong with guns faced a British force in the field. At dawn the position was found to be empty and the pursuit taken up again, the column being sighted after a 10-mile trek over the *veld*. But, at Elands Kop 20 miles further to the west, the Boers finally managed to shake off pursuit at sundown. Prinsloo, now rejoined by de Wet, doubled back in the dark and at dawn all further trace was lost. There was nothing for it but to return to the line, after the 70-mile march of the last thirty-six hours.

On 31 December Byng rejoined Elliot's columns from Kroonstadt. Only pausing to take in supplies, another week of very strenuous marching and manoeuvring followed during which the Boer force broke up into small commandos, while Mears with his Transvaalers was entrusted with the guns. Night marches followed day marches, while independent squadrons scoured far and wide. On 4 January, some 60 miles were covered with a total of 250 miles for the week's work. Marching to the neighbourhood of Lindley for supplies, news was received that de Wet was trying to work his way across the line at Roodervaal. De Lisle and Fanshawe's columns were immediately dispatched to try to cut him off. Marching light without transport, 50 miles were covered in twenty-four hours and two columns bivouacked near the line at Roodervaal, marching another 30 miles next day to meet their transport on the blockhouse line.

From Doornkloof the column moved into Kroonstadt to refit. On the march in, a Boer rode up to the advance guard scouts with a white flag and, giving as a souvenir to the scout a representation collection of Boer silver coins, also gave him a note for the adjutant who was well known by name to many of the commandos in the North East Free State. The note contained the countersigns for the week at Kroonstadt, which the local commandant thought might prove useful! The adjutant was to lunch with Major Kerr of his regiment who was Knox's staff officer. After lunch, Major Kerr suggested giving him the countersign for the day; as it was not generally given out till 5.00 pm it would save him coming back later. He had made no mention of the incident coming in and said casually that he thought he already knew it. The countersign proved correct and before General Knox could recover from his astonishment he added, 'I think I know tomorrow's too.' This also proving correct he boldly added, 'As a matter of fact I have them for the week.' Knox, always fiery, nearly had a fit.

'How did he get them? They were most confidential and only given out at retreat each evening.'

'Oh, a Boer rode with a white flag and gave them to us,' laughingly replied the officer. Needless to say, the countersign for that week was immediately altered.

It was also about this time that General Rawlinson as column commander ascending a *kopje* to look for Boers, found them there and was captured. He was popularly credited with having been relieved of his breeches and let go, as his language was too forcible to be that of an officer!

After a week's well-deserved rest the column moved out again to Reitz to take part in a move on a wide front towards Harrismith. This was preparatory to taking up their position for the first of the new methodical blockhouse drives, which was planned for the first week of February. During these operations the force found themselves at the familiar Liebenberg *vlei* near a Boer concentration camp, which had been formed in the neighbourhood. In

consequence of representations from home the policy of forcible transportation of Boer refugee families had been abandoned after the July-August drive in the Western Free State. Those families who had no farms to go to and did not wish to enter British refugee camps were permitted to form concentration camps of their own, which the British undertook not to molest and to protect from local natives, as far as lay in their power. It was near one of these concentration camps that camp was pitched early in the afternoon. A deputation came up and asked if the adjutant would visit the camp as they had a message to deliver. There was nothing to keep him in camp and he willingly went down. On arrival he found he had been deputed to be arbitrator in a dispute between de Wet and the local commandos. It appeared that after Tweefontein, de Wet had retained all the captured cavalry horses - all in the pink of condition as they had not taken part in much hard column work recently. These he now served out on payment, crediting the money to the Free State War Fund. The local commandos felt very aggrieved as they were the actual captors risking their lives freely in the attack. Finally, both they and de Wet had agreed that Captain Craufurd should be called in as arbitrator on the first opportunity as an uninterested party. After hearing the whole story his decision was that de Wet, as commander-in-chief, was within his right, but that as the Boer forces contained a large sprinkling of educated businessmen, the burghers had the right to demand a properly certified audit of accounts. Passing the Liebenberg *vlei* camp again on 31 January, he called in and asked how the decision had been received. The pact had been honourably kept but both sides, so the ladies said, were disappointed. It is, perhaps, a fair, though unwilling, tribute to the impartiality of the strange award.

Major Dyne's two months' command had not proved a success. An unfortunate incident in which he withheld fire till too late, when a party of Boers had been driven against a strong position held by the 6th MI, had led to great discontent and, unknown to their officers, a deputation of sergeants interviewed De Lisle and told him of the occurrences. Dyne, who was far from fit, was told on reaching Bethlehem, he must go for a rest into the line. The rank and file, of course, did not know this. On reaching the line a thorough rifle inspection had been ordered and during the afternoon a burst of firing broke out. On going to enquire the cause the adjutant was told the men were firing to clean their rifles. Ordering it to stop at once the adjutant went back to his tent when another burst broke out. This time the names of some half a dozen offenders were taken.

Next day being Sunday, Dyne was told he could not hold orderly room and after lunch he said goodbye and went into the line. Forthwith, in spite of the day, orderly room was held. Each delinquent was offered a court martial, which was promptly refused lest they should cause any trouble to the man they hoped would be their new permanent CO. Instead, his command was ushered in by the award to each of fourteen days' Field Punishment No.1 and discipline once more restored. Dyne very generously made a present of Souters saddlery to his successor, as well as any kit he thought would be useful to other members of the Mess. Major Craufurd continued to command the battalion till the following August when a reorganisation of mounted infantry units took place after peace and he went on leave.

By the end of January, the two great blockhouse lines of the Eastern Free State, namely from Heilbron through Frankfort and Vrede to Bothas Pass and that from Kroonstadt through Lindley and Bethlehem to Harrismith and the Natal Border, had been completed.

A further line along the Transvaal Natal railway completed the system and Kitchener was ready for his 'New Model Drives'. Elliot's two columns, with others, were consequently moved down to Bethlehem Harrismith sector to take up their jumping-off position for the new move.

With the network blockhouse lines now gradually nearing completion throughout the two republics and indeed, though to a far more limited extent, in Cape Colony itself, the long struggle was within measurable distance of reaching its inevitable conclusion.

There was to be a flicker in the Western Transvaal, where gallant de la Rey was to win two brilliant victories at Yzer Spruit on 26 February and again at Tweebosch on 7 March. The Boers, with the stubbornness of a kindred race to ourselves, were still to hold out against overwhelming odds and hopeless conditions for another three months or more before peace was finally arranged. But the new weapon – the drive in one continuous line of mounted men between blockhouse lines, strengthened by scattered infantry detachments, was, in the end, to prove irresistible.

The first great drive was planned for the first week in February. Four groups of columns co-operated under Elliot, Remington, Byng and Rawlinson

respectively. Each group was roughly 2,000 strong and was subdivided into two or more, totalling fourteen small columns in all, for facility of command. In addition, there were two small groups of columns in support, namely one under Barker in support at a kop on the Lindley – Bethlehem road, with Holmes and Marshall again in support on the left flank. The actual scope of the drive was to be a 54-mile line drawn from Frankfort to the aforementioned kop, with an ultimate pocket enclosed by Kroonstadt, Wolverhoek Junction and Heilbron railway lines. This was patrolled, in addition to the blockhouse garrisons, by armoured trains with search lights and stop columns. In order to make sure de Wet was within the drive, Byng was left within the threatened area to keep in touch, while the other three columns were placed along Harrismith - Bethlehem blockhouse line in position to wheel into line on the Liebenburg *vlei* and, if need be, face about and drive east to the Drakensberg with de Wet himself being the main object. All were in position by 1 February.

On 2 February, Byng's force came into touch with de Wet near Armstrong Drift and, in the ensuing action, recaptured the gun and pompom lost at Tweefontein on Christmas Day. Byng was now able to report de Wet within the western area of the drive and the columns were consequently kept deployed onto their line by 5 February. Shortly after daylight the great movement began, covered by a screen of scouts in one thin line from blockhouse line to blockhouse line. Then came the main body of the advanced guard of each small column; guns and transport followed a mile in the rear. The combined strength of the driving line worked out to an average of one man to 10yds frontage. At night there was to be a continuous line of picquets, practically the whole force being on outpost duty. The distance to the butts on the railway, 50 miles or so, was to be covered in three days.

Inside the area were known to be some 1,800 Boers in scattered commandos. The total forces employed in this first great drive consisted of 17,000 troops, exclusive of town garrisons, 300 blockhouses and seven armoured trains. Even so, with all this great preparation, owing to the immense distances covered, it was obvious that determined bodies of men not only could but would eventually break through but at a cost to nerves almost as severe as actual casualty lists. It was the first time in history that an advance in line had been tried on so magnificent a scale. Fortunately, the country was a comparatively easy one to manoeuvre over. If there was no hard riding or night riding, the three days' work was to prove arduous enough for all concerned and rest had to be snatched chiefly at the long midday halts. Nonetheless, for those engaged it was a thrilling experience.

Night fell on the evening of the first day with no harder work than keeping the long line intact and the long outposts stretched continuously from Doornkloof, on the south, to Holland, on the north, without a break. De Wet, once he grasped the full nature of the trap which had been set, realised that the only hope for the burghers lay in a breakout from the *kraal* in force at some point or another. Ordering a full concentration at Slangfontaine, 15 miles south of Heilbron, at midday on 7 February de Wet determined to break out the blockhouse line to the south. Only 700 burghers responded to his call, driving great herds of cattle with them. At midnight, aided by the black darkness, he struck the line at a point midway between Lindley and Kroonstadt, and a little to the West of Doornkloof, and quietly cut the fence without being detected. Shortly afterwards the cattle and their escort arrived and rushed the line by main force. Once safely through de Wet and his burghers marched on without pause for another 40 miles to the Doornberg Hills, west of Senekal, to await further news. Marshall, not far from the crossing line, with his stop column, receiving no news from the blockhouses, proved of no avail.

All through the drive on the 7th, scattered commandos had been trying at one point or another to find a gap in the driving line. Suddenly, on topping in a long sloping rise, some fifty burghers waiting and watching with their scouts on the crestline, charged through as the column came into full view. The dragoons on the right of the 6th MI were taken by surprise and the Boers were through them before they had time to shoot. The men of the 6th meantime excitedly shouting to them to draw swords and cut. Alas, from long disuse and the wet weather of winter, the swords had rusted in and would not draw, and a great opportunity was lost. The Boers got away scatheless except with one or two of their horses hit as they galloped into the blue.

The same night that de Wet broke through to the south, a combined commando under Van Collier and Van der Merwe also broke through Rawlinson's line, eventually crossing the Heilbron Frankfort blockhouse line with the loss of thirty men in each adventure. The final day of the drive was in the contracted triangle between Heilbron and Koonstadt railway and by nightfall Elliot had reached America Siding. All night long search lights of

the armoured trains played on the *veld* from the railway lines. All through the night bursts of fire broke out from one end to another of the triangle - the line of picquets in the driving line being almost stronger than that of the railway blockhouses. All were on the alert and there was no rest for anyone, excitement being at too high a pitch. A few Boers cut their way through while many were shot, but most, abandoning hope, lay down to hide until daylight when they could safely surrender. By the evening of the 8th, 286 Boers had been accounted for in killed, wounded and prisoners, besides the many unrecorded casualties of those who broke back. Of the prisoners many had been found hiding in caves or holes or even in the reed beds of the Rhenoster, buried up to their necks in water. That five-sixths of the total numbers of Boers within the trap did eventually manage to break through to liberty is a fine tribute to their pluck and determination.

No time was lost by Kitchener in organising a second drive on a still greater scale and with a more elaborate complication of moves. The new drive, which was to be carried out in three phases, was to include the whole area of the North East Free State with the Natal railway in the Transvaal as its northern boundary, the line Winburg, Bethlehem, Harrismith as its limits to the south and the Drakensberg to the east. The three phases of the operations were in all to occupy fourteen days - but not of continuous driving for all columns concerned. For this purpose, the columns were divided into two wings, starting respectively from the northern and southern flanges of the net and gradually wheeling inwards. The third and final phase was to be down a funnel formed by the Wilge and Klip Rivers with the Harrismith blockhouse line as ultimate barrier. The southern wing, mainly under Elliot's direction, consisted of six columns: De Lisle, Fanshawe, Holmes, Marshall, Lawly and DuCane. On the north were Rawlinson, Remington and Byng.

After a short four days' rest at Kroonstadt, Elliot started first on a line from Kroonstadt to Doornberg from which place de Wet had moved back into the Elands Kop area as soon as the last attempted envelopment was over. Starting on 13 February Elliot's six columns moved east to the Lindley, Bethlehem blockhouse line, which was reached on 16 February without much tangible result. On arrival at Lindley, Elliot heard that de Wet was at Elands Kop and sent Fanshawe and De Lisle on a night raid to the farm he was reported to be hiding in on 17 February. Owing to a mistake in farms given, de Wet got clear and only ten prisoners and his helio were captured. De Wet himself had, however, an unexpected escape, for 'B' Company foraying wide on the return journey saw a Boer in shirt sleeves leaving an outlying farm. A patrol of Colour Sergeant Shannon and two men went down to the farm and found only women but brought back the Boer's coat with documents inside. It proved to be that of de Wet with, among other papers, the correspondence between him and Kitchener at Paardeberg. In it was a reply to a helio message to the invested Boer *laager* from de Wet quoting the verse, 'They hemmed me in on every side I say, but in the name of the Lord I will defy them.'

Kitchener had sent, 'Those that take up the Sword shall perish by the sword.' This correspondence is now one of the treasures of a Welsh County Museum.

At Lindley a four-day halt was made for the northern or left wing to get into position. On 16 February the left wing moved from a 50-mile frontage stretching from Heidelberg to the Transvaal to Gottenburg, a small station on the Heilbron railway midway between Heilbron and Wolverhoek Junction, and drove eastward. Two days later, on 18 February, the 65 miles had been covered and the new line stretching between Standerton and Tafel Kop on the Frankfort blockhouse line reached. Two of the stop columns, those of Kier and Damant, had been drawn into the northern wing to thicken it. After a twenty-four hours' halt, the northern wing, wheeling right with Tefal Kop as pivot, faced southwards on the new alignment between Tefal Kop and the Verzamelberg by Volkerust where the frontiers of Natal, the Transvaal and the Free State met. Elliott's column in conformity had, meanwhile, wheeled eastwards and took up the line of Wilge from Majoors Drift, on the Harrismith blockhouse line, to Bachante in touch with the Frankfort line. All was now ready for the final phase: the drive between the two rivers, which was carried out by the troops of the northern wing.

By the evening of 21 February all was in readiness for a final move on the 22nd. The driving line, 60 miles long, moved forwards. Inside the corral were Steyn, de Wet and the Free State government with 3,000 fighting burghers, including 300 Transvaalers under General Alberts, whilst 30,000 men, just ten times their number, formed the troops of an ever-contracting ring. Only one point was in their favour - the scene of operations was in the most rugged and broken area of all the Free State. Before the week was out would come

Majuba day, now ominous as the anniversary of the first great surrender at Paardeberg two years before. Desperate situations demand desperate measures and on 23 February de Wet, finding himself at the head of 900 burghers, albeit encumbered with a great mob of frightened refugees, decided to try to break through. The British driving line had reached the line of the Hol Spruit and a full moon made easy the location of the picquet lines to enterprising scouts. A weak spot at the junction between Byng and Remington was chosen and the attack launched at 11.00 pm. Garrett's New Zealanders put up a desperate resistance but the thinly held line had no reserves. In the end, de Wet with 600 stalwarts broke through, carrying some of the horde of refugees with them. In doing so he left fourteen dead on the field and twenty wounded who got through with him. The New Zealanders had twenty-three killed and forty-three wounded in the vain attempt to hold the gap. Other small commandos managed to dribble through, either by crossing away from the known drifts on the Wilge or in the wild foothold hills of the Drakensberg. Foiled in two attempts to break through Elliot's line, on the morning of 27 February, Meyer and his commando of 570 men surrendered to Rawlinson. The total captures in the drive resulted in 778 prisoners, fifty Boers killed, 25,000 head of cattle, 2,000 horses and 250 vehicles. This was the biggest single coup since Prinsloo's surrender in July 1900. The drive had, however, far more effect than mere numbers of capture. Shortly after this date the wives and mothers who had so far cheered their men to resistance through every change of fortune now tentatively, and for the first time, talked of peace to their acquaintances within the columns.

De la Rey might still snatch victory in the Transvaal and resistance and negotiations be prolonged another three months but, with the home front pierced, the end was now definitely in sight. The fight at Langverwacht was thus to usher in the peace, so long awaited.

Little rest was allowed by Kitchener either to his weary troops or to their still more weary quarry. On 4 March the columns moved out again to take up their position for the third drive. Like its predecessor it was to consist of three phases. The opening moves were to be carried out by two wings moving out first independently as before. The actual drive itself was to cover the same ground as in the first drive: from Liebenberg *vlei* to the Kroonstadt railway; but first the great quadrilateral Lindley, Frankfort, Tafel Kop, Harrismith was to be swept so as to drive the Boers as far as possible into the corral. Twelve columns were to take part, a slight reorganisation having taken place during the halt, including Byng's return home to England. This time Elliot's group was to form the northern wing. This moved first, marching northward on a 50-mile front from the Harrismith to the Frankfort blockhouse line. Nine columns, amounting in all to 10,000 men, took part in the manoeuvre. As soon as Elliott had departed, three columns under Rawlinson, consisting in all of 5,500 men operating in rear of Elliot's march, swept northwest on a 30-mile front between Majoors Drift at the junction of the Molen and Wilge and a point on the Lindley blockhouse line south of a kop on the Lindley – Bethlehem road. On reaching the northern blockhouse line Elliot's line from Tafel Kop to Bothas Pass was to turn about and, pivoting on Tafel Kop, wheel on to the new line of Liebenberg *vlei*. This Elliot did on 9 March, but the day following the driving line was weakened by the withdrawal of four columns to the Transvaal in consequence of Methuen's disaster at Tweebosch. The attenuated line was no longer equal to its task and many Boers broke through in the dark.

Mentz, with 400 men of the Heilbron commando, was, however, still within the net. The driving line now reduced, pressed forward on the 60-mile frontage between Lindley and Frankfort and, in two rapid marches, reached the railway on 13 March with a total result of some 100 Boers killed, wounded and prisoners. On the last night Mentz galloped the line north of Heilbron and broke the wire for a distance of 60yds by sheer weight of impetus and passed out within 100yds of a blockhouse. Another lot had similarly burst through the Kroonstadt Lindley line the same night.

It was on the last night of this drive that a lieutenant of the Wiltshires, but recently joined, whilst visiting his outposts from in front was unfortunately challenged and shot dead through the heart, not having apparently heard the challenge in the pouring rain. Only one shot had been fired and it was only when he was missed and a search made, that the tragedy was discovered, his body lying only a few yards from the sentry post.

For fully a month the troops of the North East Free State had been driving backwards and forwards between the blockhouses with only the briefest halts for supplies, and the constant work, day and night, was beginning to tell. A week's halt was consequently given at Kroonstadt before the next drive started during which many changes were made in Elliot's own two columns. De Lisle was transferred to the Klerksdorp District to take over charge of a newly formed Australian column and Little of the 9th Lancers took over the cavalry column De Lisle had been commanding for the past three months. With the departure of their column commander, the 6th MI, De Lisle's own regiment which had always been under his command since its first formation two years before were now transferred from their cavalry brigade to Fanshawe's column of Yeomanry and colonials, in place of the South Australians, who in turn were being transferred to one of the new Australian columns.

The new drive, which started on 19 March, was to be a straightforward one, east between the northern and southern blockhouse lines, and was to move at a speed which would allow rest and methodical searching of the ground covered. Five groups of columns were to take part in order from right to left: Elliot, Barker, Nixon, Garrett and Remington; some 14,000 troops in all were on the line Heilbron, Doornkloof in position for the drive right across to the Drakensberg.

Moving through an empty country the Liebenberg *vlei* was reached without incident on 23 March. Experience had shown it was in the latter stages that the Boers tried to break out in force so, in consequence, Garrett was sent on ahead with his column to concentrate on breaking up any gathering against the driving line. At this juncture, a week or more of heavy rain set in which ended by upsetting the whole scheme. Garrett pushed on over the *vlei* and on to the Wilge at Streydpoort Drift and managed to get one of his columns, that of Gordon, over the Drift. However, owing to a sudden spate found the two others, those of Dunlop and White, cut off behind. Meanwhile, the main driving force, after managing to get across the *vlei,* was, in turn, held up by the Wilge. There was thus one small column, that of Gordon, on the east of the Wilge exposed to Manie Botha and 1,500 Free Staters if he ventured to move away from the shelter of his column's guns and the whole of the driving line immovable between two rivers in flood. It was not until 28 March that Dunlop was even able to get a cable across and so pass supplies to Gordon from his transport. Not till 1 April did the Wilge sufficiently subside for the driving force to cross, Remington having moved back to Frankfort to cross the rivers by the bridge at the junction near the town.

The rain still continued with little intermission and, on 2 April, at midnight Manie Botha broke through. Every post was on the lookout as prisoners had given warning of the attempt. From 9.00 pm onwards many scouts tried first one post and then another along the line. At last, at 1.00 am, in the pitch dark and pouring rain, having located the British picquets along a wire field fence, suddenly a Boer bugle rang out and a mingled mob of frightened cattle and shouting horsemen crashed onto the picquet line. In the darkness and confusion some hundreds of horsemen galloped through Garrett's line. On 7 April the drive ended, producing no very great material result. Just ten Boers had been killed, seventy-six captured, with 4,800 cattle and horses collected after daylight from the charging mob. Some 180 vehicles and three derelict Krupp guns were found by Nixon in the Leibenberg *vlei*. Elliot then moved on to Harrismith to refit.

Rumours of peace parleys were in the air while Kitchener's immediate attention was concentrated on bringing the Transvaal to a proper realisation of the situation. Elliot was consequently left without any definite instructions and, after a short, three-day halt, moved back again into the driving area with no special objective in view. Any thoughts of immediate peace were, however, somewhat rudely shattered by a determined attack on Fanshawe's column at Rietkuil by 300 Boers on 14 April. The Boers finally made off into the mist where it was impossible to follow. For the 6th MI this was to be almost the last skirmish, for while the peace parleys of 12 April at GHQ had broken down, delegates' orders were meanwhile being sent to various commandants for a final discussion in the line.

Lindley was reached on 21 April and, after filling up, Elliott's force moved via Bethlehem and Retiefsnek into the Brandwater where some 1,500 Boers had taken refuge from the drives. On the way down de Wet came in on pass moving to one of the Boer conferences. He accepted lunch with the 6th MI but this was not allowed and the officers met him at tea instead. He then stated that only the wide-flung flanks with which De Lisle's column marched had saved the British from attack on more than one occasion. In and around the Brandwater, as indeed elsewhere, the situation was curious. There was no armistice, but messengers between commandos with duly authorised passes were allowed to pass freely and notified conferences were not broken up. The columns received orders not to go out of their way to raid or attack Boers. While the Boers regarded the situation as an armistice, the British held that the Boers must keep out of their path to avoid capture.

The first two days in the basin, the columns found themselves gathering protesting prisoners in most satisfactory numbers. Finally, practically a whole small commando was captured in a farm. The Boers sent in a sporting young lady riding side saddle on a man's saddle with the off stirrup thrown over her knee, and with white flag to protest. With a Cape Town high school education, she was fully equal to the occasion and opened her protest with, 'I like you officers cheek, capturing prisoners during an armistice.' The respective views were explained and it was agreed to put the question up to Kitchener - the prisoners remaining compulsory guests meanwhile. For two nights there was a sing song for the guests, at the end of which the Boers sang 'God save the King' against whom they were still in arms. On the third day came orders for release. As the rearguard moved off, back came a messenger with a white flag to thank the British for the hospitality and confirmed there would be no sniping that day.

There remained one more drive, to take place in the first ten days of May. Getting into position, Barker's column found Manie Botha wounded in a farm. Their openly expressed pleasure that he was merely a casualty prisoner and not dead, as rumoured, must have been a full compensation for not retaining his liberty to the end. For once Elliott was not in the drive but his columns formed as it were the "guns" along the line of Butts. Bruce Hamilton from the Transvaal first drew up his troops on the line between Vereeninging and Greylingstadt, then drove south with seven columns. On 3 and 4 May he drove to the blockhouse line and formed up on a front of 60 miles from Kopjes Station to Frankfort. 200 prisoners had been taken. Lining up from his new position, the 40-mile drive to Elliot on the southern blockhouse line was carried out on 6 May in one day. By noon, Elliot's line had been forced by two separate charges, first against the 3rd Dragoons, and two hours later against the 7th Dragoons. Mentz with his commando of 300 surrendered. Two days later, in a return drive on 9 May to the Frankfort line, only twenty-two burghers were captured.

The following day, on 10 May, a definite armistice was proclaimed. Negotiations dragged on for another three weeks while the delegates at Vereeninging strove hard to stave off the actual admission of defeat.

This interval had been spent quietly between Kroonstadt and Lindley, very largely in polo, gymkhanas and shooting competitions. The practice for Proclamation Parade had not been without its humour when a solitary Boer sniper from a hill within range which had not been picqueted, put in one shot of protest onto the parade ground with a spectacular effect not provided for in the programme.

A few days later, after peace had been declared, a small party of two officers and three other ranks, all but the officer interpreter of the ASC, set out with three waggons and seven days' supplies for fifty-one families around. The first farms had been fully warned and the reception was cordial. The next farm or two were truculent and, on leaving, the party were not sorry to receive seven surrendered rifles and bandoliers with which they promptly armed themselves. At the next farm a big gathering had arranged a lunch. Songs were sung including the favourite war song which ran, 'Bobs is in Pretoria but where is Lady Bobs?' Lady Bobs referred to the big Long Tom captured at Helvetia in 1900. Urged to describe the new king, the officers pooled their unused stamps and handed them around. Then a dance and 'God save the King'. At the last farm visited the reception was distinctly hostile. A party of young armed Boers rode up and claimed the whole turnout as captured, saying they knew nothing of peace and demanded the officers should strip. The surrendered rifles were loaded and both sides stood fingering their triggers till the women intervened. The farm was left with full military precautions - the young burghers being warned that an unprovoked attack would count as highway robbery with murder. Two parting shots were fired as the party rode off but as they went wide no further notice was taken.

After the middle of June Fanshawe's column moved into Kroonstadt. First Elliot and then Fanshawe said goodbye; the 6th MI came under Knox as normal garrison as all columns were being rapidly broken up. In contrast to the constant moves of the past two years life was now quiet enough. Before the regiment settled down to garrison duty, they had distinguished themselves by sending down a waggon and fatigue party in view of all who might care to see, to cut the acacia wood from Kroonstadt's sacred park, which had so far avoided all scars of war! If wood rations had often been small its energetic collection, where available, had hitherto been a praiseworthy act.

For a mounted corps a wide radius of ground for a shoot was always sought but with peace now declared, permission had to be obtained. The best shooting *vlei* belonged to a Boer, notorious among his fellows for refusing any concession. The officers meant to shoot at the *vlei* and there was likely to be trouble, so taking the bull by the horns, so to speak, the major rode straight through to the Boer and said that he had heard that the old gentleman was most tenacious of every right but that taken in the right way and treated with diffidence, no one in the country was more reasonable and generous - and might we shoot? The old gentleman was frankly taken back at this reputation, admitted, however, that it was true though seldom realised and gave permission. In the end, greatly to the surprise of the Dutch neighbours, quite a cheery friendship sprang up with the 6th MI officers.

General E. Knox was not long in following Elliott and Fanshawe home. In column days the 6th MI had often been a thorn in his flesh when in for a rest. They rode him off the ball at polo more than they rode at the ball itself and their conduct at the theatre was more cheery than orthodox. The mounted columns generally rode in as if they had bought the town. He had not forgotten the forestalling of his countersigns and there was the recent episode of the park. All this he pointed out to the OC of the regiment at a parting lunch and then handed over a really generous hamper, or rather a Cape cart, full of his latest stores including one of his famous buzzard cakes, with crème de menthe to wash it down. Shortly afterwards, on 31 May, the two Boer governments put their signatures to the formal surrender and on 1 June, 'Peace' was flashed along the wires of the country.

Change was in the air and at the beginning of August, Major Duncan of the Royal Scots arrived to take over command. The 6th MI were out at the time, strung out on detachment holding the line, 110 miles long, across the Free State from Bothaville to a kop on the Lindley – Bethlehem road to hand over to the South African constabulary. Major Duncan was to take over a revised 6th MI and, though the name lingered for many months, it was no longer the 6th MI who had fought and ridden for the past two and half years over such a large portion of the British South African empire.

For all their raids and forays, long rides by night and dashing skirmishes by day, for all their many hours in the saddle, it was the blockhouse line, in the end, that won. On that long line of blockhouses then, farewell.

Chapter 10

Training - School of instruction for mounted infantry and manoeuvres

Up until the late 1880s the training provided for mounted infantry work had been conducted on an ad hoc basis in the areas they were to operate in. This followed the line of impermanence and improvisation that applied to the organisation as a whole. The question facing the future of the MI was one of necessity and how it would be organised. There continued to be much professional opposition to a permanent, mounted infantry, particularly from the cavalry who saw the MI as a threat, interfering with their role as the eyes and ears of the army, while the infantry were chary of a system that would deprive battalions of some of their best officers and men.

These differences were discussed in the 1886 lecture given by Major E.T.H. Hutton, chaired by Lord Wolseley. Hutton opened the lecture by acknowledging that all agreed on the general utility of the mounted infantry; however, opinions differed regarding the necessity of such a force and, if it was, how it was to be organised. In his view the principal object of mounted infantry was firstly to provide an improvised substitute for an expensive cavalry in small and hastily organised expeditions and, secondly, in the case of a campaign on a large scale, to furnish a force of selected infantry sufficiently mobile to act in conjunction with cavalry. He agreed that a permanent force would tend to deteriorate into an inferior brand of cavalry and that it was only in time of war that a mobile force of this description would be of value. Hutton saw that a training school was essential, through which a selected number of infantrymen should pass with the object of being formed into a separate arm in an emergency. This, he believed, would, 'be best and most readily obtained by a school and cadre corps of mounted infantry to be established at Aldershot'. Lord Wolseley and Sir Redver Buller, among others, warmly endorsed the scheme and, despite opposition from the Treasury, the formation of a school and cadre of mounted infantry at Aldershot became a reality, opening in 1888 under Hutton's command with Captain E.A.H. ('Puntie') Alderson as his adjutant. However, like the MI itself, the establishment of the school was on an improvised and opportunistic basis.

'Curly' Hutton was then Deputy Assistant Adjutant General on the headquarters staff of the Aldershot command. As commandant of the School of Instruction for Mounted Infantry was graded as DAA & QMG and, besides commanding the mounted infantry school, he had also to supervise the School of Army Cooking, a test indeed for the versatility of the British officer. Hutton is generally, and justifiably, known as the father of the mounted infantry. He was heavily involved with obtaining official recognition of the arm and initiating the scheme of training which bore such notable results in small campaigns including those in Rhodesia and Somaliland and culminating in the Second Anglo-Boer War. A rifleman, he was described by General Sir A. Godley as a 'very forceful character' and by Major Tomlin as 'one of the liveliest livewires, abounding vitality and with a manner which endeared him to all, he found no difficulty in imparting his own enthusiasm to those under him'. These characteristics were required to overcome the initial obstruction met from the cavalry lobby opposition, particularly from the Inspector General of Cavalry, Sir Drury Lowe, who was reluctant to allow the use of cavalry horses during the furlough season, the indoor riding facilities and cavalry instructors to teach equitation. Hutton temporarily overcame the shortage of horses by suggesting that mounts of regiments due for service in India be used. The problem was finally settled in 1891 when 120 cobs were purchased for MI training purposes. Hutton collected together sufficient men and mounts to form a company and the first training course of the mounted infantry regiment began in January 1888.

The training programme's objective was to train officers and other ranks in all that was necessary for the care of horses, riding and stable management, the movement of mounted troops in formation and in the field including the advance and withdrawal, patrols, outposts, observation and extensive reconnaissance. Initially the course was to last two months but was subsequently increased to three months.

For the first three days the training consisted of instruction in the riding school and manège, given entirely by the company officers, with occasional help from a cavalry sergeant and two troopers allotted to each company. On the fourth day the companies were paraded, mounted, without arms and, despite a large and critical audience of cavalrymen, were put through their infantry drill. The whole affair was an unqualified success, so much so that General Sir Drury Lowe, who was watching from a distance, was convinced that the drill had been squadron and not a company. At the end of two months HRH the Duke of Cambridge, accompanied by the adjutant general and Sir Redver Buller, inspected the result of this experiment and was more than satisfied. Thus, during the following fourteen months no fewer than forty-two officers and 1,100 NCOs and men were trained to form a nucleus of mounted infantry, complete for service in time of war, at an estimated cost of £1,725.

On 21 March 1889, a second lecture was given by Major Hutton with Sir Redver Buller in the chair, to a larger and representative audience. Sir Drury Lowe generously set the seal of success upon the enterprise saying: 'Cavalrymen have been asked to lead the way in discussion. I have listened attentively to the lecture and, as a cavalryman, I cannot see that there is anything I can discuss in a combative spirit. We have heard a great deal about mounted infantry, but to my mind the lecturer has disposed of one great argument against it by avoiding any advocacy of their interfering with cavalry. I feel convinced that the use of horses or ponies by infantry as a means of locomotion to enable them to keep up with cavalry will be an enormous advantage to a cavalry force. I do not see that Major Hutton has taken upon himself to interfere with the role of cavalry at all. I will not admit that mounted infantry can take the work of cavalry but, in many cases, they will be an enormous help to cavalry. They might save them from a great deal of hard work in a standing camp, on patrol on roads and such like. As the lecturer has not inferred that mounted infantry will interfere with cavalry in their duties, I cordially agree with him and sympathise with the movement, and trust I may be allowed to remark that as we have the advantage of the quartermaster-general's presence this evening, he will use his influence with the Secretary of State that in organising and training such a force, other branches of the service shall not be interfered with.'

The training of the mounted infantry regiment was, in the following winter, renewed and an additional four companies were trained, thereby forming a nucleus of a mounted infantry regiment for the Second Army Corps. The system rapidly gained in popularity among infantry regiments and the cavalry were gradually released from the duty (to which all cavalry officers were not favourable) of lending their horses. A yearly allowance for cobs entitled the training school to be independent of horses on loan.

Godley described the average MI officer as, 'One of those ready for adventure, who did not believe that the be-all and end-all of an officer's life was a round of duty in barracks with alleviations in amusements, whether hunting in Ireland, or tea parties at Southsea, in accordance with the station at which they served and with their individual tastes.' Tomlin added "most came from families where the horse was not looked on with the jaundiced eye of the Psalmist, who recorded that, in his opinion, the horse is a 'vain thing for safety'. Where our purses allowed it we kept a horse, though it may only have been of the class of those in the livery stable of Charlie Wise, which were said to be hunted four days a week and to have a rest in the dog-cart on Sunday.

Major Tomlin relates that when he was with the mounted infantry on Salisbury plain in 1903, he was dining one night with the GOC Evelyn Wood. They talked of the horse habit and agreed that it was the best habit on which to spend what little spare cash one had. The field marshal dug into the past and said that when he was serving with an infantry regiment (for his career had taken him from the Royal Navy to cavalry and infantry regiments) he had only one horse. How many days' hunting did Tomlin think he had during the months of January and February? Of the fifty-nine days available, Tomlin calculated that excluding Sundays, it was possible to hunt, and assuming that the meets were near, a maximum of fifty days. Evelyn Wood's answer was 'forty-five'.

Year by year, the different companies were trained at Aldershot, establishing a strong body of mounted infantry within the army, who could at any time be mobilised and formed into separate units.

Other duties called the original commandant and adjutant, but the school continued under their successors. Hutton was succeeded by Lieutenant Colonel 'Bobby' Auld, Northumberland Fusiliers. Auld was not a mounted infantry man, but General Sir A. Godley records, 'He was a most charming and efficient commanding officer and in Captain St. George Henry of his own regiment he had an adjutant who was the ideal of all that a mounted infantry man should be. A first-rate horseman and horsemaster, a first rate man to hounds, a good sportsman, a good disciplinarian with a thorough knowledge of soldiering and a charming personality, he inspired us all.'

Godley related his early MI experiences before he succeeded as commandant of the School of Instruction for Mounted Infantry at Aldershot in the spring of 1901 on his return from South Africa: 'I was fortunate, in 1894, to be attached to a light infantry company of mounted infantry commanded by Captain Thomas D'Oyly Snow of the Somerset Light Infantry. His subaltern's included H.L. Ruck-Keene of the Oxfordshire Light Infantry and 'Smiler' Brooke of the Kings Own Yorkshire Light Infantry. We were quartered in Malplaquet Barracks in the North Camp with a mess and stables nearby. Other officers beside myself, were attached for instruction, among them Henry Wilson of the Rifle Brigade and Harry Rawlinson of the Coldstream Guards. Two of the permanent staff, well known to all MI officers those days, were Regimental Sergeant Major Rouse[17] and the orderly room Sergeant Nesbitt, both ardent upholders of the MI gospel. Nesbitt, when asked by a would-be detractor, why young officers who had done a course seldom came back replied, 'Of course they don't. They almost invariably become adjutant of their battalions.' The training was strenuous and the hours long, as it was a point of honour with all MI officers to begin with attendance at early morning stables which, on a raw, cold winter morning by the light of stable lanterns, was a veritable test of keenness.'

The course lasted for three months. Auld was in due course succeeded by Lieutenant Colonel Honourable Frederick Stopford of the Grenadier Guards and, in 1895, Godley was appointed adjutant (St. George Henry having gone to the Egyptian army) and continued the training of successive companies without intermission until 1896, when the special service battalion of MI for service in South Africa[18] was formed. Captain T.T. Macan of the Cameronians succeeded Godley as adjutant.

In 1898 two companies, one mounted on the one company establishment of cobs and the other on cavalry remounts, were raised to take part in cavalry manoeuvres. Godley was appointed to command them, and they were encamped on Hounslow Heath for training, eventually marching to Salisbury

[17] Served in the Ashanti expedition, Boer War, awarded D.C.M. and the Great War where he was commissioned as lieutenant and quarter master in the Machine Gun Corps and awarded the M.C.

[18] The Mashonaland Field force commanded by Colonel Alderson.

Plain. There they were attached to the cavalry brigade commanded by Colonel French of which the regiments were the composite regiment of Household Cavalry, one squadron each from the Blues (1st and 2nd Life Guards) and the Royals (1st Royal Dragoons), and the 10th Hussars. The Cavalry Division was commanded by General Sir George Luck.

The training was, as a rule, simple and strenuous. Sir George was evidently a great believer in drill and plenty of it. Starting out from Sling Plantation, where they were camped (close to what is now Bulford camp), to a rendezvous just across the Avon, a favourite day's work would be as Godley describes: 'The division was formed up with one brigade in front, one on each flank (one of these being echeloned behind the other), with the horse artillery, mounted infantry and massed machine guns in the centre. Heading for Devizes, we would advance, first at a walk, then at a trot, gradually increasing the pace till the whole division was at full gallop. On arrival at Lavington Down, a distance little short of 10 miles, the order would be to turn about, and the same manoeuvre was repeated back to camp. Luck himself would gallop in an out of the ranks, and the whole performance was certainly a test of endurance for both man and horse, especially the latter. It was all very well for the officers who could have two horses out and, as at a drag-hunt, have their second horses waiting for them at the check! I always did. But it was trying the troop horses very high. The Household Cavalry horses, then of much inferior quality to what they are now, and fresh from guard duty in London, or mild drill in Windsor Park, suffered especially, and it was said that a trainload of fresh horses from their regiments had to be sent to replace casualties. Our stout little MI cobs stood it the best. Anyhow, rumours of the mortality and of the condition of the horses reached the ears of the Commander-in-Chief, Lord Wolseley, and he decided to come down and see them for himself. This he did and, on a very hot day, stood for hours, with his back to Sling Plantation, while all the horses of the division filed past him - a good performance on the part of the great Field Marshal, who was by no means a young man. The result was a considerable curtailment of our activities. From Salisbury Plain we marched to Blandford and formed part of the force under General Sir Redvers Buller, which from there manoeuvred against the Duke of Connaught and his Aldershot Command.'

Auld, Stopford, Godley, Macan are names familiar to the trainees before the Boer War in which the mounted infantry came into its own. With the outbreak of the South African war in 1899, Stopford passed on the command of the School of Instruction for Mounted Infantry, Aldershot to McCann. After returning from service in South Africa, in the spring of 1901, Godley was appointed to succeed McCann as commander. He was responsible for training MI detachments for service in South Africa where the war was in its guerrilla phase, with its heavy reliance on mounted troops to defeat the enemy. On his appointment there was one Gordon Highlander and two composite companies from militia units and provisional battalions in training, and these were the first to be dispatched for service during the last year of the South African war. Many other companies followed and, for the first time, the Guards Brigade sent a company, commanded by Ted Trotter (Grenadiers). This was followed by a second company under command of John Ponsonby (Coldstream). W.B. Bridge, Highland Light Infantry, was adjutant and was succeeded by W.E. Lawrence, South Wales Borderers, who went to Longmoor and was succeeded by Captain H. Findlay (Buffs). At the same time the mounted infantry schools were made responsible for the training of some composite regiments of Yeomanry for service in South Africa. By the final stages of the war some 16,000 men had seen service across twenty-eight MI battalions under the direction of Sir Ian Hamilton. 'Puntie' Alderson, who had gone to South Africa in 1899 as CO of the 1st Battalion Mounted Infantry, was the inspector general.

At the end of the Boer war a considerable force was left in South Africa, which included mounted infantry formed by the different regiments in the command. These acted as a training school, as their establishment had to be made up by recruits when the trained soldiers gradually went to the Army Reserve.

Following their performance in the South African operations, mounted infantry was all the rage, allowing the best field officers to be selected to command battalions; company commanders and officers were among the best their regiment could provide. The officers selected to be trained required at least two years' service, have completed a trained soldier's course of musketry, have some knowledge of horsemanship and be able to ride.

In 1903, it was decided to form three schools of instructions, at Longmoor Camp in the Aldershot command, the second at Bulford Camp on Salisbury Plain, and the third at Kilworth Camp in the south of Ireland. This would enable four battalions to be trained: two at Aldershot and one each at Bulford and Kilworth. The War Office, in a letter to Godley dated 20 September 1904, explained that the establishment of the training schools contemplated the training of a company for each battalion of infantry in the United Kingdom, with certain exceptions. It was thought that in the event of a large force of MI being required for some special purpose, such as a fresh rising in South Africa, it would be possible to mobilise sixteen battalions of MI from the trained companies. The equipment for these sixteen battalions would be held at Aldershot and Salisbury Plain, six sets each, and the remaining four sets at Kilworth. At the time of writing the War Office stated they were unsure how far the accumulation of equipment at the three places had progressed and further advised that no decision had been taken how the MI would be organised if sent out in large numbers. Lord Kitchener had laid down that it was to depend on the nature of the campaign.

Godley was appointed commandant of the Longmoor School. The camp of wooden huts had been established during the Boer War, and many additions, particularly in the way of stables, and alterations had to be made. Major Maurice Tomlin who had served with the MI during the war in South Africa was at the start of Bulford Camp on Salisbury Plain that was under the command of Colonel Rolt. He described the origins of Bulford as an example of haphazard methods which often produced good results. Guy Bainbridge (later Major General, KCB) was appointed to command at Kilworth.

Kilworth did not have a happy start and by summer 1904 it looked as if it would be temporarily suspended in September. Vivian Henry (on the staff at Kilworth) informed Colonel Godley in July that in his opinion the pertinent issues were:

> Accommodation - it was exceedingly difficult to find available accommodation that was in a satisfactory state of repair to stabling, barracks or training ground.
>
> There was a personality clash between Guy Bainbridge and the DAAG Cork District, William Hickie (later Major General Sir William Hickie, KCB). According to Henry this was due to Hickie's 'insane jealousy of B. He has continually gone on influencing Pole Carew (GOC 8th Division based in Southern Ireland) and Johnson against him simply because he hates him.'

> Henry believed that Grenfell[19] was also influenced as his remarks on the GOC's report went along the lines that the report was satisfactory but, as there appears to be a difficulty about getting sufficient men to volunteer, the school does not seem to be quite a happy family. Henry refutes this saying that the statistics for the last three companies trained showed that out of a total intake of 491 men there were three non-volunteers and although he had not the statistics for the fourth company, he thought all had volunteered.

Godley had corresponded with Guy S. St. Aubyn, who was on Lord Grenfell's staff, who advised that they were unaware of the friction existing at Cork and this knowledge explained several points they could not understand. Aubyn continued that Lord Grenfell fully realised it would be a pity to suspend MI training, even temporarily for the winter, but the question of accommodation for men and ponies during the winter months was as yet, unsolved. He stated that, 'The staff over here allege that no place exists once Ballincollig is re-occupied by the cavalry. The Government should harden their hearts and build places for the MI. As regards locality, the Curragh, of course, suggests itself at once and the ideal arrangement would be to treat the MI as corps troops, like you do at Aldershot; but the practical difficulty which arises is this, that at Aldershot your divisional generals have no territorial areas or responsibilities, whereas in Ireland these are considerable. Distances are great, units scattered, and consequently greater latitude has to be given to GOC's of districts than is necessary, and, I may add, desirable. In addition, at Aldershot, corps troops are directly and permanently under the eye of the Army corps commander, whereas in Ireland they are apt, owing to the abovementioned reasons, to get out of his sight and it would be difficult, though not impossible, to prevent the GOC of the district having a say in their management.'

Lord Grenfell went on to discuss this matter with General Stopford, who was director of military training at Horseguards. St Aubyn assured Godley that he could rely on him to let Lord Grenfell see the MI side of the question

---

[19] Later Field Marshal **Francis Wallace Grenfell, 1st Baron Grenfell**, GCB, GCMG, PC (Ire) (29 April 1841 – 27 January 1925); at this time he was C-in-C Ireland.

and that though there would be difficulties he believed they could be overcome. However, he closed with the remark, 'But it seems to be agreed on by all hands that Bainbridge's appointment was an unfortunate one.'

The courses were to be made up of volunteers picked by their battalion commanders who were trained infantry men with a minimum of two years' service, be of good character and physique, not to be above a certain weight and be a first-class marksman. The officer was required to be at least a fair rider and have some knowledge of horse-mastership.

Stable parades were conducted under the supervision of the officers; it was important that the cobs were well looked after and fed, for them to be able to stand up to the demands of the strenuous training agenda, namely daily parades, field days involving many hours in the saddle covering significant distances. In addition, they were expected to be available for up to three days a week, for the officers and NCOs' use, playing polo in the summer and hunting in the winter. Hunting was a recognised part of the training for the officers and a few selected non-commissioned officers, and gave them unrivalled opportunities of acquiring knowledge of, and eye for, the country. Typically, the training programme would look like:

Week(s)	Activity
One and two	Personal equitation.
Three to eight	In order: i. Section; ii. Company; iii. Battalion mounted and dismounted duties and roles.
Nine to eleven	Tactical exercises.
Twelve	Field firing and scouting competitions.

Schools had been established outside the United Kingdom in Egypt, Malta, South Africa and India. In India these were at Poona, Amballa, Bangalore, Sialkot and Fatehgarh. British infantry were trained during the winter months and Indian infantry in summer. Douglas Haig was the inspector of cavalry in India from 1903 to 1906 and he is seen as a heavy influence on the decision to close the schools. This took effect from 1908 with three schools closing and ceasing to train Indian infantry in the mounted infantry role. Bulford, Ireland and Malta schools closed in 1906 and, by 1912, the only school operating was at Longmoor, which was considered large enough to cover the MI needs. In 1912,

Major Hope raised the question in Parliament of how many companies were trained on each course at Longmoor and how many will be trained in the future following the transfer of 450 training cobs (out of about 880) over to the cavalry and following the return of cavalry regiments from South Africa. Colonel Seely, Secretary of State for War, informed him that, during 1912, six companies were trained per course and there had been three courses. This would be reduced in the next year to two companies per course and three courses to be held. Either the Secretary of State was not correctly briefed or was possibly being disingenuous as the School of Instruction for Mounted Infantry, Longmoor was closed in 1912.

General Sir A. Godley, in his autobiography, summarised his thoughts on the benefits of mounted infantry training: 'Up to that time there is no doubt that the MI training had been a most valuable asset to the army. It had widened the horizon and broadened the tactical view for many of the best infantry officers of the army. Its example had exercised a very wholesome effect on the cavalry and yeomanry and had made them realise that dismounted action and good shooting were as necessary to the role cavalry as the *arme blanche*. It had also provided a nucleus of men sufficiently trained in mounted work to be a valuable auxiliary to cavalry, and to act as mounted troops in circumstances where no cavalry were available, or where the conditions were more suited to the use of troops made mobile by the use of ponies, camels or other means of conveyance, than to cavalry on horses, trained primarily for shock action and the use of the *arme blanche*.'

Manoeuvres and joint exercises were an important measurement of the effectiveness of the training and readiness of the army to fulfil its role in time of hostilities. Following the South African conflict these were held more regularly than prior to the war, and a number of opportunities were made available (in 1904, 1905, 1906, 1909 and 1912) for the mounted infantry to mobilise and participate in joint manoeuvres, inspections and cavalry training exercises.

The 1905 manoeuvres were held at Churn, during the period 27 to 29 June. Lieutenant-Colonel Fanshawe, Oxford Light Infantry, was invited by Colonel Godley to observe the MI concentration and cavalry manoeuvres. He wrote up his observations on 30 June and sent them to Godley; an edited version following provides an insight into the MI on manoeuvres:

'Each battalion had its own camp and officers' mess managed by A&N Stores at a rate of 12/6d per day. There were two canteens between the five battalions. The sergeants shared one large mess.

'Sections and companies coming from Longmoor kept their cobs in the lines of the battalion in which they were mobilised. All the other cobs were kept together on the lines of two battalions but were freely lent to all battalions on evening of 27 and on the 28 June so that each company might have a company drill and riding instruction.

'In the companies I saw at the drill, the men seemed at home on the cobs, but I thought the officers were not keeping a steady enough pace and trotted too fast. This seemed to me a mistake as regards instruction to the men and hard on the cobs who nearly all were out twice that day (a subsequent note was made saying, "some of them were out 4 and 5 times and did about 10 hours work.")

'The men mounted on horses other than MI cobs on the 29th were formed up clear of their camps to take over the horses of the 1st Cavalry Brigade, RE, Artillery and some ASC. At 6.30 am to facilitate handing the horses back to the proper squadron, the men were told to remember the place they drew their horses from and to take them back in the evening to the same place and in the same order.

'The horses were handed over saddled, with rifle buckets, nose bags and feed, water bucket and one other came with sword frog. Only snaffles were issued. Possibly a few of the horses would have been better with the bit as well, but undoubtably the majority were best without the bits. A subsequent note was made stating, "Some had bits – all COs were asked to put bits on any they thought would be unsafe without." I think it would have been better on this occasion if the surcingles had been sent and if the stirrup leathers had been arranged in the usual MI way without the brass runner. I think greatcoat straps should have been sent and coats and capes carried.

'Judging from the way some of the machine gun and pom-poms were used in the first phase the orders could hardly have been sufficiently explained by battalion OCs or attended to and studied by some officers. On arrival at the canal, battalion OCs and other officers concerned (especially pom-pom and machine gun officers) and those of the leading company hardly seemed to realise the importance of using their spare time in making themselves acquainted with the ground just in front of them over which they were about to fight. Some very good reports came in from the reconnaissance, but one report came very late – the orderly having apparently gone the wrong way. This was likely due to either not selecting the right man to carry the message, or to the sender not instructing the orderly with sufficient care as to his way.

'I don't think that on becoming engaged with the enemy every unit reported the fact at once, stating on what line they themselves were and on what line the enemy were, and whether the latter were holding or giving ground. Rifle firing can seldom be heard over a mile and often not nearly so far, so all units must not rely on the sound of their fire and the enemy's fire keeping their superiors informed of how the fight is progressing.

'The concentration was quickly done. Did the OC of each unit on approaching the concentration rendezvous ride on himself or send his adjutant on to get orders?

'Units were hardly led at a slow enough pace to prevent horses and men getting a little out of hand and to ensure everyone being closed up in his place. I noticed this in the 6th Battalion advancing to the holding attack. I don't think scouts were used enough here. In the main attack, after dismounting, some of the officers and men did not use the ground and work quite intelligently according to the ground. The horses appeared to be left well concealed.

'The getting back to the horses when operations were over and concentrating for the march past was very quickly done and was one of the most remarkable performances of the day.

> 'I don't think it can be too much impressed upon officers that they should dismount and lead frequently whenever the time is available and the place suitable.'

Lieutenant Colonel Fanshaw finished by stating: 'The whole show was I should think, a great success for the MI.' The Inspector General of Cavalry, Major General Robert Baden Powell, wrote in his *Notes on Manoeuvres at Churn 1905* that this had been, 'On the whole a satisfactory camp of exercise.' The inspector general listed observed faults (often noted by Lieutenant Colonel Fanshawe) with its effect and noted remedies, summarised as follows:

> 'General fault was want of knowledge on part of line leaders of exact methods of moving. Instruction needed in cavalry training to this effect.
>
> 'Great tendency to get too wide in distance and interval: consequently, dissemination of the division.
>
> 'A tendency also to get into wide, unwieldy lines eg line of squadron columns – before it was necessary. "He wins fights who manoeuvres last."
>
> 'Want of quickness in carrying out any change owing to overuse of gallopers instead of signal.
>
> 'Instruction and subsequent drill-practice in the principles of attack on the various arms would have been valuable, before such themes were attempted in the field as manoeuvre, but owing to circumstances this was not possible at Churn.
>
> 'More certain arrangements necessary for signalling inter-communication between units when scattered – a system wants perfecting. Some days signalling was satisfactory, but often it was not.
>
> 'A great improvement is necessary in the accuracy of leading both in the small units as well as in the greater ones. Minute accuracy of direction is

essential to effective drill and cannot be got where the leader is shifting his line or the men are looking at their dressing. Direction is best gained by having a trained guide out to the front of the line where everyone can see him and regulate their direction accordingly.

'Pace is the other great key to good order, and pace was not satisfactory as a rule. The trot was nearly always too fast. Supporting lines nearly always have to move at a somewhat faster pace than the leading line, and if this is going at a fast trot the others have to be galloping to keep their relative places. Then, if in addition to this, the front line is guided in even a slightly wavy direction (as was almost invariably the case at these drills) supporting lines are constantly either checking or galloping all they know to keep up.

'In some squadrons, though not in most, there was great stringing out in passing through narrow defiles such as gates, gaps in brush etc, the men tearing along singly to regain their places after getting through. Troop leaders should always wait till their troop is complete after passing such a place and then bring it up at increased pace but in good order to rejoin the squadron.

'Dismounting at every opportunity was not carried out at all satisfactorily. Squadron commanders should be expected to order it on their own initiative, the moment that there is a halt – providing that no time is lost in mounting when the brigade has to move on.

'Rifles must invariably be taken off the saddle when men dismount, in order to relieve the horse.

'Off saddling ought to be carried out once every two hours – if only for a minute. Several rubs which I saw could never have gone so far had this been practised. This off saddling ought to be practised on every mounted drill of the squadron or regiment, in order to make it habitual with all ranks.

'Rifle action dismounted and fire discipline should be more under the section leaders. In certain regiments squadron leaders were trying to direct the whole fire which is bad practice.

'The skyline is still too much trespassed on by staff, scouts and formed bodies. Three regiments I noticed particularly good at keeping off the skyline, but others were too careless in the matter, and it might mean everything to them in war.

'The "regrettable incidents" occurred in charging unreconnoitred positions, and one brigade exposing itself to annihilation under heavy artillery and machine gun fire with little chance of escape.'

Baden Powell, however, recognised much that was satisfactory:

'Brigade drill, though loose, was very much better than I expected seeing that the regiments had had no opportunity of practice at it – and many had not even been able to get regimental drill owing to want of ground at their stations.

'Conforming by line leaders to the general scheme of movements was very intelligently and quickly carried out on most occasions. It is a point to be remembered by leaders of lessor units. Be on the lookout to conform, or to back up where you see you are wanted or to be in best position to help in case of being wanted or, if nothing presents itself, join the reserve. Never be found with nothing particular to do and doing it.

'Scouting by individual scouts and officers' patrols was very satisfactory. I hope scout officers will in the training of scouts, develop combat-scouting (which includes rapid and accurate transmission of important verbal reports).

'Signalling individually was well done – and occasionally inter communication all over the field was very good – but as I have said

> before this should be made a certainty by well devised schemes on the part of the brigade signalling officers.
>
> 'The condition of the horses was satisfactory. Horse mastership has made real progress and can be yet improved in small points such as I have suggested above.
>
> 'The efficient working of the mounted infantry was especially commendable.'

The report concludes with Baden Powell's suggestions which focused on dissemination of the day's operations and lessons drawn to the troops in a form of precis highlighting good work. He used as an example of 'Erskine's bluffing manoeuvre of his section of guns.' This, he believed, would be instructive and stimulating for the men and, in any event, he would require officers to explain the day's operations. He also considered that it would be advantageous for NCO troop leaders to attend all pow-wows and participate in tactical war games of troops in contact.

Chapter 11

MI swansong and Russian campaign 1919

Following the Boer War, a Royal Commission was established to, 'Inquire into the military preparations for the war in South Africa, and into the supply of men, ammunition, equipment, and transport by sea and land in connection with the campaign, and into the military operations up to the occupation of Pretoria,' otherwise known as the Elgin Commission. By taking oral statements from senior officers involved, it provided a chance for the supporters and detractors of the mounted infantry to put forward and reiterate their views and differences. Influential supporters of the MI included Lord Roberts, Lord Wolseley, Sir Evelyn Wood and Sir E.T.H. Hutton. Their opponents, principally cavalry officers, were Douglas Haig, Michael Rimington and John French (who was to a lesser extent opposed, but was prepared to offer limited, qualified support). Tensions between the cavalry and mounted infantry had been around since the early days. The cavalry's primary concern was that of the mounted infantry assuming more of their traditional roles and duties. They saw this threat as detrimental to their future security and ultimately, as a possible threat to their very existence.

The MI position had always been that they would complement the cavalry. They would relieve them of certain basic roles and duties; for example, the MI would provide the mounted element for expeditions where there were no available cavalry units. Taking into consideration the acknowledged numerical insufficiency of cavalry in proportion to other arms, these duties assumed by the MI would free up their resources enabling them to concentrate on their traditional roles. It was recognised and acknowledged that the MI would not be able to provide the *arme blanche* role but, with their dismounted fighting tactics and use of the superior infantry rifles, they would provide better support and stronger firepower than the cavalry could with their carbines. Hutton in his lecture of 1889[20] claimed that the functions assigned to the MI would not detract from the necessity of the cavalry. Indeed the three arms – infantry, cavalry and artillery – had their distinct and well-defined duties to carry out. He further stated, 'I am thus anxious to anticipate any statement that it is the intention for mounted infantry to usurp the functions of the cavalry more than do the functions of the horse artillery by the action of their attendant machine-guns.' Lord Wolseley did not regard MI as, nor want them as, a substitute for cavalry. He suggested that, 'Armed with a long rifle, cavalry could perform the duties of mounted infantry. However, it would be similar to employing a Nasmyth hammer to break walnuts.' Despite these reassurances that their role was to work in conjunction with the cavalry and take on some of their less arduous and more mundane roles, there was always a level of professional jealousy that was never fully resolved and was bought to the fore in the South African war and its aftermath.

The South African war had been one requiring mobility; the cavalry operated largely in a counterinsurgency role with few opportunities to use *arme blanche* tactics. The mounted infantry had finished the war in a strong position with some argument that the rifle and its firepower had displaced the cavalry and their edged weapons as the significant force on the battlefield. However, it was the mounted infantry that was disbanded in 1913 and would take no part as a formation in the Great War. The cavalry during this period had used the time to reform and reorganise itself such that it was included in the BEF order of battle that went to France in August 1914 but its role, after the initial war of movement, proved limited. Indeed, the advances in technology made to meet operational requirements led to greater use of the internal combustion engine in its various guises, to provide mechanised means of mobility and offensive firepower.

The MI's demise was caused by several factors that turned against them over the decade following the Boer War. Perhaps the most important was the decision to retain the MI on an impermanent basis. Hutton had, from the start, argued for impermanence; he saw that it had many advantages, while the

[20] "mounted infantry – Its present and its future" by Major E.T.H. Hutton, D.A.A.G. presented on 21 March 1889; Major-General Sir Redver H. Buller, V.C., K.C.B., K.C.M.G., quartermaster general to the forces in the Chair.

danger of a permanent force acting with the cavalry would lead to it becoming a very indifferent cavalry. The temptation existed for the infantry soldier to drop the infantry function in favour of the more glamorous cavalry role.

Maintaining an MI organisational capability on an impermanence basis, relying on abstraction of trained MI men from their infantry battalions when required, was less costly. Estimates ranged between £109,000 and £178,000, while the cost of raising additional MI battalions on a permanent basis was estimated at between £390,000 and £472,000. In addition, the expense of equipping a mounted infantryman was considerably less than that of a cavalryman in creating a mounted force capability. Hutton estimated the cost of cavalry horses in 1886 at £70 for a 5-year-old (the minimum age required) and he had sourced Arab ponies for the MI contingent used in the Suakin expedition of 1885, at a cost of £26 each. Cobs and ponies were more suited mounts for the MI soldier, particularly in close and difficult country where to carry out his duties he needed to be able to quickly and easily mount and dismount. A cavalry trooper required a sturdy mount that could bear a heavily accoutred man for the *arme blanche* role. Training requirements of the MI were of a short three-month duration concentrating on equitation and horse management skills, the fighting skills having already been ingrained into the soldier. Cavalry troopers and his horse required between eighteen months and three years to become fully trained.

Organisational impermanence by relying on abstraction of trained soldiers at the time of mobilisation together with the MI tradition of improvisation had its disadvantages. The lack of an overarching doctrine for the MI left them without a defined method for deployment and role on the battlefield. Who should they operate with, infantry, cavalry or independently? Although their role seemed assured following the Boer War there was no single organisation to identify lessons learned and apply them to a doctrine giving guidance for future operations. Without a defined doctrine they were reliant on interpretation of the various published mounted infantry manuals and regulations, in particular those published between 1897 and 1909, with the *Mounted Infantry Training 1909* coming nearest to recording their functions as it focused on their training, capability and operational roles and less on stabling duties. Lacking formal guidance, commanders had a degree of flexibility that resulted in differences in training methods, tactics and, to some extent, command.

As the decade progressed a number of the MI's original, senior supporters were losing their influence while, since the end of the Boer War, the cavalry had used their time wisely reforming and reorganising themselves. They recognised the need for mobile firepower. For defence they replaced the carbine with the Lee-Enfield rifle for use in dismounted action. For offensive capability the machine gun was introduced (before the MI), in addition to their cold steel capability. The introduction of cyclist infantry sections further contributed to the MI's demise, as this form of transport also provided a means of swiftly and efficiently moving bodies of infantry. Though it was acknowledged that the limitation of the bicycle did not give them the freedom of movement cross-country that a horse or cob could.

Functions that had been identified for the mounted infantry were being fulfilled by cavalry regiments repatriated from overseas stations where permanent postings were no longer required. As an example, the MI had been responsible for performing a divisional mounted troop role by allocating a company to an infantry division; these were being replaced by cavalry squadrons supported by a company of cyclists. Mixed mounted brigades was an idea considered, but not implemented despite deployment during the 1905 manoeuvres. Although joint exercises were held there was no formal integration or defined co-operation between the two arms. Their detractors, in particular Douglas Haig, actively denigrated the mounted infantry and criticised the organisation for failure to maintain equitation skills as refresher training was not common. According to Godley, Douglas Haig never rested till shortly before the Great War when he succeeded in getting the schools broken up and the whole idea of MI training abandoned.

Support for the MI was not guaranteed from all infantry officers; abstraction of trained MI soldiers from infantry battalions at the point of mobilisation was unpopular with battalion COs. They were reluctant to lose well-trained men at this time when all experienced soldiers would be required to help integrate reservists to bring the battalion up to strength. It was also argued that the abstraction of these experienced soldiers was detrimental to discipline and *esprit de corps*, as their behaviour and influence provided a standard that young soldiers could aspire to.

The disbandment of the mounted infantry became a reality in 1913. On 8 August 1913 a meeting of the Army Council discussed the reorganisation of mounted troops for the British Expeditionary Force, and the fate of the mounted infantry was sealed. The Army Council was concerned by the impermanence basis of the MI. It believed that there would be difficulties in mustering sufficient numbers, using the system of abstraction, to mobilise the required number of companies. The question of forming a body of reservists to make up any shortfall was objected to on the basis that these reservists would lack recent training and may be a danger to the unit when facing modern European armies with up-to-date equipment and trained in the latest tactics. In addition, the council members were concerned that abstraction weakened infantry battalions as they would then become reliant on reservists as replacements. In their opinion the reformed cavalry could now fulfil both a defensive role, applying their musketry training, and an offensive role as shock troops using their hard-edged weaponry. Other MI functions could be covered with the support of cyclist units. The result was that the mounted infantry was considered redundant and, with the closure of the last school of instruction for mounted infantry the previous year, there was no source available for the supply of trained recruits.

The Russian campaign of 1919

After the first Battle of the Marne in the First World War, neither side were again able to supply cavalry with any effect on the Western Front. Only in the open country of Palestine was it possible to make any effective use of cavalry. The question of mounted infantry never arose. But, after the Armistice of 1918, in the operations in North Russia in 1919 a small force of mounted infantry was formed.

The terrain in North Russia was ideally suited to the use of cavalry but, unfortunately, there was none available at that time. It was therefore found necessary to form mounted infantry units to cover the rear of the retreating force, which consisted of 4,500 men under Brigadier General Sadlier-Jackson, composed of the 45th and 46th Royal Fusiliers, a company from 201st Machine Gun Battalion and two companies of Royal Engineers.

The brigade was operating on the River Dwina, with Archangel as its base, and had advanced 400 miles upstream. It was then found that the Bolsheviks were cutting the lines of communication in several places and that mobile troops needed to be trained to deal with this menace. Owing to the thick forest, horses were the only solution. Volunteers were called for and a large number of men responded to the call. Many of these dropped out when the mounts made their appearance – hardy Russian ponies, minus saddles and bridles. The makeshift use of blanket and rope respectively called for considerable skill. However, a fair number survived the test and, after much hard work from all ranks, an efficient mounted infantry corps came into being.

They did valuable work in keeping the enemy at a distance until it was found that retreat was inevitable. It was then that their supreme value was demonstrated. Private W. Hartley of the 45th Royal Fusiliers recorded the following account of mounted infantry work in the retreat to Archangel during the early part of September 1919: 'On 6 September, a large force of Bolsheviks were reported entering the village of Ivanovshaya on the left bank of the Dwina and were a menace to our retreat on Seltso. A party of twenty mounted fusiliers and one platoon were organised to attack at dawn on the 7th. The latter was strongly resisted until the mounted men charged with their bayonets and drove the enemy off into the forest, taking forty-five prisoners and killing twenty-seven. At least sixty were wounded, according to prisoners' accounts. This so demoralised the enemy that two battalions coming up took flight and fled.

'On the morning of 8 September, another patrol of mounted fusiliers spotted a column of 300 Bolsheviks making a flank movement. Thereupon the officer sent for a platoon of foot and two units attacked at the point of the bayonet. This time not only were eighty-one "Bolos" killed and ninety-nine captured, but three guns were also taken. Our own casualties were very slight.

'The "Bolos" were pressing us heavily now, always being just where we had left. Therefore, the mounties were in the saddle continuously for days, having to make sorties from the rear in order to drive the enemy back a few *versts* (roughly 0.66 miles), allowing the main body to retreat a little further. On the night of 9 September, a heavy barrage was opened on the village of Gvanovshaya which was held by twenty mounted fusiliers under Lieutenant Green (nicknamed Green's Horse). At about eight o'clock two battalions of Finnish Red Guards attacked the village from the north, south and west. Soon Green's Horse were entirely cut off and were compelled to resort to hand to hand fighting for over two hours before they were able to rejoin the main body, with the loss of two killed and eight wounded and missing.

'By now the evacuation was making good progress and the force was embarking on the barges for Archangel. The Mounties had served their turn and were dismounted. There were many touching scenes at the parting from their faithful ponies who had served them so well and whose second sense had saved many a life and brought many a lost man safely out of the desolate forest.'